Writing a novel is a lonely expedition. We set out with this story that irritates the mind, but it doesn't come to fruition without help.

To my ride or die: Elissa Juliette. The amount of love and inside jokes I could put here, would be more than the word count of this book. Same page, babe. Same bloody page.

Thank you to the ones who read this story when it was a mere figment of my imagination: Kriszta and Baiba. You pushed this story into the direction it needed to be.

Thank you to author friends who inspired me to continue and always believed in me. Especially, Bethany Olin, KB Benson, and Courtney Millecam. You reminded me this story was important and only I could write it.

Thank you to Chrissie, Matt, and Cindy for ripping apart the little details of this shabby story and making it that much better.

Thank you to Rachel George for a stunning cover illustration. You made Lume (and Seclus) real.

Thank you to @chicklen.doodle for making trading card character art I always wanted.

Thank you to my beta readers who jumped in to provide their incredible feedback: Crystal, Erin, Hailey, Heidi, LaNona, and Lauren.

Thank you to Rachel Schade, for not only being an amazing author and guide, but also for being an awesome friend. Thank you for pouring over each word and helping me feel better about choosing to publish.

Thank you to everyone who loved this story from short snippets and small fragments. This wouldn't have been finished without the unending love and encouragement.

And most of all, thank YOU, for picking up this book and giving it a chance.

To my GAMERS: no matter which stage you enter, just jump and shoot and you'll make it through. I love you more than you will ever know.

Book Cover Illustrated by Rachel George Illustration
Title Page Illustrated by Rachel George Illustration
www.rachelgeorgeillustration.com

Map created by MGG

AISN (ebook) B09X8QMWLR • ISBN (paperback) 979-8-9867457-0-1

LUME

TM Ghent

for Willow
and for the ones who find comfort
in the darker side of books –

"Promise you'll say that I made a lovely villain
when you tell them your side of our horror story."
- ang frank

Author Note

Orson Scott Card wrote an introduction when a new release of *Ender's Game* was published. He wrote something that resonated with me long before I decided to become a writer. I am going to share it with you here:

> This is the essence of the transaction between storyteller and audience. The "true" story is not the one that exists in my mind; it is *certainly* not the written words on the bound paper that you hold in your hands. The story in my mind is nothing but a hope; the text of the story is the tool I created in order to try to make that hope a reality. The story itself, the true story, is the one that the audience members create in their minds, guided and shaped by my text, but then transformed, elucidated, expanded, edited, and clarified by their own experience, their own desires, their own hopes and fears.
>
> The story...is not this book, though it has that title emblazoned on it. The story is the one that you and I will construct together in your memory. If the story means anything to you at all, then when you remember it afterward, think of it, not as something I created, but rather as something that we made together.[1]

Remember as you turn these pages, I am but a small part of this. I have created the structure of the world, Lume. I've built the basic frames of these characters, and you will witness it as it unfolds. But you, the reader, get to complete them.

---

[1] Orson Scott Card, *Ender's Game* (New York: A Tor Book, 1991), xxv-xxvi (with omissions).

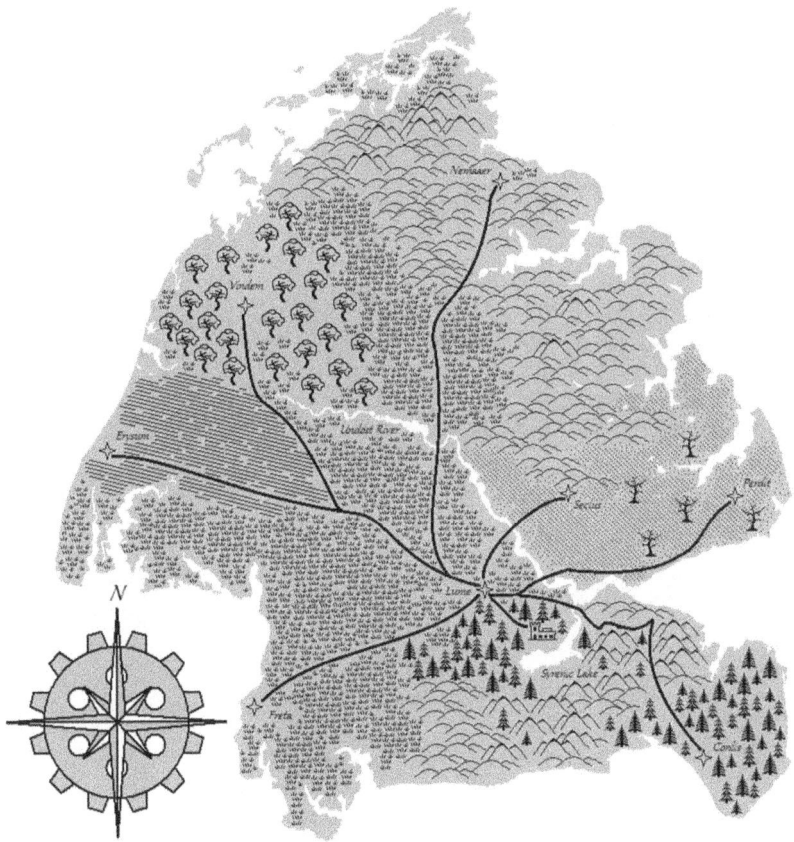

Time

Prince Rajveer Klauduisz tapped the metal fingertips of his left hand as he pulled in a long drag of his cigarette, staring across the castle grounds before him. Fiedel's rays, as it started to slip below the horizon, warmed him beneath his emerald jacket. The full Petram, a mirror and faint shadow, followed on the other side of the sky.

He let the smoke escape out of his nose as a horse-drawn carriage slowed outside of the front gate. Pinching the cigarette, he took the final pull before snuffing it out beneath his shoe.

"Sir?" One of the Satelles, Jude, looked at him, and Rajveer nodded. Jude walked to the carriage to open the door and let out their newly arrived guest.

Rajveer adjusted the neck of his jacket, forcing it closed with the metal loop. Although internally he wanted to do none of this, there were certain expectations to be upheld, even if everyone at this point was smart enough to see through the act.

But this was the everlasting game he and his father, the king, would play from now until one of them died. Fates, Rajveer was borderline impatient waiting for him to go already. The King offered him the throne on one condition: Rajveer was to be married. Marriage wasn't something Rajveer could agree to, especially not after his first engagement.

So, instead, he was forced to have a guest from the capital city visit their vast and lonely castle every petrik. And with Petram reaching its fullness again, the women, and the occasional man, would attend to keep the gossip alive. But unbeknownst to his citizens, they were paid for their silence.

Rajveer struggled to remember the name of the woman attending today, even though he saw it when he gave his approval for her invitation to the castle. She entered the grounds now, her peach gossamer dress dragging along the stone walkway.

She placed her hand over her heart and bowed before him. While her eyes were focused on the ground, Rajveer turned to the Satelle Captain, Amicus, willing him to understand the hidden question. Amicus chuckled silently before mouthing the name to him.

When the woman rose, Rajveer put out his left hand, grasped hers, and folded it into the crook of his arm before moving to enter the castle.

"It is a pleasure, Emilia, for you to join us." Rajveer slipped the façade in place as he prepared himself to face his father.

"Prince Rajveer, the pleasure is all mine." Her light blue eyes danced with what appeared to be genuine appreciation. He walked her up a handful of entry steps and directed her forward to the dining area, when she slowed in an attempt to look in the direction of the grand room beyond. Luckily the curtains were pulled closed, leaving more to her imagination as Rajveer was not inclined to want to give her a tour.

They entered the dining hall and Rajveer kept his eyes trained on her so he could witness her curiosity. It was one of the few delights he looked forward to with every new visitor. Her blonde curls fell behind her shoulder as she gazed at the ostentatious painting of the late queen, his mother.

At the far end of the room, his father rose from his seat, gesturing for Emilia to take the one next to him. With an air of elegance, she walked the length of the room, the lights from the chandelier seeming to move with her.

Staff members were spurred into action by her presence. Some pulled out chairs, some poured wine, and one with far too much enthusiasm informed the chef the dishes were ready for service.

Rajveer trailed at a slower pace, claiming his own chair before immediately accepting the offered glass, watching as the brilliant golden color swirled. He drained it quickly, letting it fill him, anticipating the eventual blurring of the edges of his reality.

King Richard took no time before beginning his list of never-ending questions. Emilia was charismatic enough to keep up with the ceaseless conversation. They spoke of her parents' standing within the temple system, her own education coming from the temple itself instead of the local school, and of course, trade and politics of the crown.

Rajveer nursed his refilled glass, the honey and spice an amusing battle in his mouth, as the discussion continued. He didn't mind the idea of politics, the necessity to have discussions among others and the citizens. It was understood and appreciated. But for the last five fiedations, his father refused any dialogue with him at all, resorting to "you wouldn't know" or "you wouldn't understand." Wasn't an only heir— raised by books hand selected by the king, trained by the Satelle Captain, and required to shadow the king on various diplomatic issues— supposed to be the person who might possibly know?

Rajveer had ideas, grand and vast and compelling ideas, to push the capital and their cities upward to more wealth and grandeur. He wanted to bring Seclus, the newly created underground city, under their reign, which meant possibilities for new technology and better advancements. His ideas, he thought, were better than ignoring the city's existence and letting the Lawless take control. He wished his father didn't stomp out every new vision or difference of opinion. To his father, Rajveer's ideas were like throwing a lit match on wet kindling.

King Richard quieted for a moment. It was jarring to Rajveer's ears after they were numb from the ceaseless babble. The staff's shoes tapping along the floor echoed in the silence as they decorated the oak table with plates filled with wild rice, cheese stuffed mushrooms, candied carrots, and roasted duck with a sticky glaze.

Rajveer stared at the food before him, letting the aroma meet his nose, when he noticed movement in his periphery. Emilia brought her glass upward, the bubbles shifting erratically with the movement before toasting to the king's longevity. Rajveer kept his composure and did not immediately roll his eyes in response. Continuing their political discussion, they both began devouring their meals, while Rajveer began his own. The cheese was tangy, a bright contrast from the earthy mushrooms, but with each bite, his mind still wandered.

After the queen's death, the king had locked himself away in the castle. However, even with the king's lack of diplomacy, the capital continued to prosper from the long-standing history of receiving a substantial amount of wealth and local goods from the cities on the continent. The king ruled over them all, until a new city sprouted relatively close to the capital. His father presented opinions and directives regarding practically every issue, except for Seclus. On the ideas of how to address this new city, the king always found a way to remain silent.

Rajveer interrupted their conversation, not sure exactly what they were rambling about. "What would you do about Seclus?"

He leaned back into the chair, glass in hand, while discarding his napkin onto the nearly cleaned plate as he alternated his stare between the king and Emilia, both shocked.

4

The former narrowed his eyes and, as far as Rajveer could tell, clenched his teeth. He might have even heard the grinding if not for the groaning of the wooden chair when he shifted positions. The latter held onto her manners, dabbing her napkin near the seams of her mouth as she swallowed her last bites of food. "I beg your pardon, Prince; what did you ask?"

"Seclus. You know, the city of Lawless, as the citizens call it. What would you do with it? I mean, you were talking about politics and giving your opinion on other crown matters, so I was curious what you thought about Seclus."

"I'm not..." Emilia hesitated a moment, looking between both the king and the prince. "I'm not sure I am in position to give such opinions."

"Please, indulge me." Rajveer raised his glass to his lips, taking another sip of the wine.

Rajveer was certain Emilia's neck was going to break with the amount of swiveling it did back and forth between himself and his father.

"Erm." She hesitated as she reached for her glass. Rajveer imagined she did so to buy herself time before continuing.

"What are you getting at, boy?" The king's words were coated with disdain. *Boy?* His father must be losing his edge to blurt such a minor assault at him. Turning his attention back to Emilia, he continued, "You don't need to answer that. Clearly, the prince has forgotten his place and his manners."

"Have I, father? You two had no issue discussing every other aspect of the crown's politics. It seemed an appropriate time to discuss the real issue facing the continent."

"You would be wise to keep your mouth shut, Prince, before I find a permanent solution for you."

The snicker escaping Rajveer's lips couldn't be helped. "The same solution you gave to me before, dear father?"

Rajveer quickly glanced down at his hand before meeting his father's eyes once more.

"That's enough," the king growled. His annoyance was tangible enough, Rajveer could have sworn the table shook.

Rajveer knew when to stop pushing, so he did, allowing himself to go back to his wine. But he knew the real issue would never be addressed. Emilia looked down, pushing grains of rice along the porcelain as if it were the most important thing in her life.

Rajveer, unsure he could deal with anymore groveling, drained the last of his glass before the chatter could begin. He rose from his seat and tucked a hand into his pant pocket, while gesturing his arm to Emilia. She shifted awkwardly to wipe her face and rise from her seat, before curtsying to the king and thanking him for the delicious meal and great entertainment.

Emilia wove her arm into Rajveer's and they both moved from the dining hall. He could feel the tension and possible anticipation rolling off her in waves. He dropped her arm, allowing her to climb the stairs to the second floor before him.

As they walked the hall, her fingers caught on his right hand as she swung her arms next to him. His entire body tensed and he forced himself to continue walking. Rajveer chastised himself for allowing her to get on his right side, blaming his own lack of attentiveness during his haste to leave dinner. He shoved his right hand into his pocket as they continued forward to his chambers.

Rounding the last corner, Rajveer noticed Amicus headed in their direction from the other end of the hall. He gestured for Emilia to enter his chambers.

"Go ahead and get ready, I'll be there in a few minutes."

Within seconds, she pulled his face to hers, her tongue exploring his mouth, tasting of wine. She sucked on his bottom lip as she pulled away from him slowly. Rajveer only stood there, hand on the door handle, as she wiped her mouth with her fingers and winked before turning into the room. Rajveer slowly pulled the door to a close as he turned back to the Satelle Captain.

"Having fun, sir?" Amicus teased.

Rajveer forced a small chuckle, thankful his childhood friend was able to see the humor in every situation. Unfortunately, no amount of it could save his relationship with his father. He took a deep breath, calming the waves in his stomach in preparation for what he was about to say.

"I think it's time, Amicus."

"Time? As in..."

"As in, my time has come. We can't delay this any longer."

"Are you sure this isn't the wine speaking, Raj?"

"Yes. Something needs to be done."

Rajveer clenched his teeth as he began massaging his temples from the headache teasing there. These false courtships he was forced to continue in order to provide hope to his citizens were proof enough changes needed to be made with his father.

"Yes, sir. I will be sure to have Danika take care of getting the message delivered immediately."

"Tomorrow is a new day."

"It is, sir. Do you require anything else before you retire?"

"Yes. The lady would like more wine."

"Yes, sir."

Rajveer began opening the door to his chambers, squinting into the darkness. The room with his bed remained unlit, dark with possibilities and wild fantasies, all of which wouldn't happen here with Emilia. They couldn't. Because for as much as Rajveer longed to have intimacy, he was anything but indecent. Not when his heart had been stolen away in the dead of night fiedations ago.

Escape

Theodora hated many things: crowds of people, dresses, snoring, people who slurped while drinking. But right now, what she hated most was warm weather. It forced a sticky layer to form across her skin and a tightness in her chest as she struggled to catch her breath. Worse was the knowledge there was no way to escape it.

She thought the warm days were over with the season changing from solta to astrum. But no, today remained a blistering day. Earlier, Fiedel had barely risen before little drops of sweat covered Theodora's body as she laid, staring at the ceiling, pleading the fates to reconsider the heat.

But they didn't. The fates either didn't hear or blatantly ignored her, which given their history, was entirely plausible.

After a day filled with various jobs around the capital city, Lume, she now stood on the street overlooking the market. Luckily, Fiedel's descent out of the sky had begun, hopefully allowing for the heat to fade as well. She brought her arm upward, blocking the light emanating from the golden orb.

The market square was overrun with people, all bustling between each other and vendor carts, which bordered a massive marble fountain erected in the middle. Encircling the wide area were shop buildings made of lackluster stones, their maintenance long forgotten, the faces of the buildings having lost most of the decorative jewels tucked within. They stood in stark contrast to the daily vendor carts which arrived in their brilliant shades of greens and blues, vibrant against the bleached stones.

After a full day, including another job at Reme Temple, Theodora had hoped her late arrival to the market would have allowed her to escape such throngs. And yet, the heat hadn't stopped the Lumens from their congregation at the square, like vultures to a carcass.

Theodora tugged her locks into an unflattering bun, fastening it with a leather cord. When she dropped her hands, loose strands fell into her face and against her neck, rebelling against the containment. She sighed and began her march along the cobblestone streets.

Every day was market day now. As a child, it was only every petrik. The horses would pull the wooden carts into the square, and she would toddle around with the other local children, throwing pebbles into cobblestone cracks while her parents fixed dumguns and sold flash crystals. The other vendors would sell their own goods.

Today? A walk through the market was an onslaught of senses. The number of vendors attempting to earn more klaud grew exponentially with each fiedation. Yes, such was the luxury of the king, as with his predecessors before him, to name the land's currency after himself.

Theodora walked through the maze of carts, their colorful merchant banners hanging limp on their poles waiting for the fates to grace them with a breeze. The smell of food tickled her nose, making her stomach turn in delight: turkey legs, bread, sticky sweets, and pickled vegetables. She remembered the first time she tried a pickled beet; how it had stained her fingertips red when she tossed it into her mouth, the way her mother had thrown her head back in laughter at the sight of Theodora as she squinted her lips closed to the bitterness. Her father's eyes sparkled over her mother's shoulder. Pure unconditional love had radiated from that moment.

However, none of these traditional scents did anything to mask the underlying sweat of the other patrons as shoulders were forced to brush against one another.

The memories of her parents were not something she allowed herself to think of often. After the amount of time that had passed since their deaths, she should have known better than to allow her grief to surface, especially at such a public location. Her hand gravitated to her hip, deftly searching for her sheathed dagger. The cut of the sapphires decorating the metal dug into the pads of her fingers, a sensation that she found helped ground her and focus her thoughts.

She continued her meandering throughout the shops in search of klaud and distraction. The vivid yellow banner of Amabel's Sweets came into view, and it propelled Theodora's steps forward with the prospect of friendship. She spotted the woman's bright cloth wrapped around her head, today a vibrant jade, and waited until Amabel's eyes located hers.

Theodora pulled klaud from one of the pouches around her waist, flashing them from behind the various patrons crowded at the cart. Amabel nodded, spotting them. *The usual.* Baked dark chocolate with coarse sea salt sprinkled on top. The salt was traded to the capital from the seaport, Freta, a good three days of travel from them by horse. It cost Theodora more than some of the other offered delicacies, but it was well worth it.

Amabel reached into the wooden cart to pull out the chocolate, wrapping it in waxed paper as patrons pushed and shoved for her attention. She left her apprentice alone and moved over to the buildings, jerking her head in the direction for Theodora to follow. They met up against a building face, the stone scraping against Theodora's bare upper arm as they leaned into each other, the wall providing a little relief as the coolness seeped into her skin.

Theodora took the chocolate. It was soft under her fingers, starting to melt. She took a moment to thank Amabel and a moment longer to appreciate the woman. Her darkened brow glistened with sweat, but nothing could wipe a smile from her face.

Amabel tugged at her canary dress, more chocolate smearing from where her fingertips stretched the fabric.

"Why did you gesture for me, Amabel? It looked awfully busy around your cart, and I don't mean to take you away from your business."

"I've heard word of a possible new task from Hakon. He is staying at his home in Seclus for the next couple days until he can find someone to volunteer. I don't know the details, but the rumors are of his dead mother and possible inheritance. It's supposed to be good klaud."

Shop owners always found the best gossip. Although some things were exaggerated or mistaken, a lot of truth ran from the mouths of customers.

"Thanks for the information, Am. I appreciate it." She pulled the waxed paper open and broke off another piece of chocolate. The sticky sweet left behind its mark on her fingertips as she slid the piece into her mouth. A small moan escaped her lips. "Fates, this is good. No person should be this good in a kitchen."

Amabel's chuckle shook through her shoulders, but it was cut short when a loud crash interrupted the bustle of the market. Theodora noticed the commotion surrounding Amabel's cart, patrons pushing and pulling at each other, curses and insults being thrown as quickly as punches, before Amabel rushed for her apprentice.

"Go find Hakon!" Amabel shouted over her shoulder to Theodora before pushing her large frame through the swarm of bodies, throwing free samples in hopes to calm the crowd before Satelles arrived.

∴ ∵ ∴

As she headed north, Theodora's mind raced with why Hakon's task couldn't find interest, but also whether she could make it to Seclus in time. The journey alone for tools or a quick visit for supplies took most of the day. And now a possible task? Learning the details and politics? She looked to the sky, gauging the time she had left.

The account of how Seclus came into existence was vague, blurred, as if someone had attempted to erase it from the pages of history. It was the conflict of their time, for as long as Theodora could remember. Seclusians had somehow appeared in their world. One day there was nothing. Petram rose and fell with the darkness, and the next day, Seclus was an underground city on the outskirts of the capital. It was entirely illogical, but they had no other information.

With a protective king and his overbearing Satelles, it was remarkable the city was able to thrive. Seclus appeared to be positioned strategically, far enough to not immediately be considered a threat but close enough to have influence. With a knack for tinkering, the Seclusians brought to society their shineguns and more reactive flash crystals. They learned to fix and improve Lumen weapons. They adapted so quickly and fiercely, Lumens started thinking of Seclus as not only a city, but an underground capital.

Word spread. From Freta to Vindem to Conlis, the power of the capital would soon be shared, and yet nothing came down from the crown. There was no invasion, no squashing of the gossip. Nothing except the nickname of the

Lawless city, Seclus, which was entirely nonsense because Lawless had always plagued the Lumen streets.

Theodora herself had even been labeled one of the Lawless, given her willing nature to commit tasks brought to her by her fellow citizens, but she only thought of it as a way to earn klaud. In her youth, she had focused on easier barters: running goods back and forth for the vendors, sweeping steps for the elderly, picking herbs for the temple. With each fiedation, her work expanded to stealing back goods, disposing of bodies, and poisoning traitors.

The intense chattering of people died away as she left the capital boundaries and headed for Kadena's outskirt stable. The buildings started to thin, these appearing to be in even greater distress than those in the heart of the city. Fields and trees filled in the space, cobblestones transformed to green, and the sounds of people were replaced with those of birds and insects.

Although Hakon's task nagged her, an incomplete item on her list, she knew she needed to wait. She had a limited amount of fiedelight left, and a trip to Seclus, even by horse, would take time. The labyrinth of the underground was so vast she wouldn't make it to Hakon's before dinner. And to interrupt a man during his meal? It would be careless.

The stable came into view and Theodora was mildly grateful to delay Hakon's task. She missed the openness away from Lume, with nothing but unlimited freedom and time. The Undost river snaked along the backside of the fields, the rolling hills giving way to the dense foliage to her left. King Klauduisz' castle gleamed brightly in the late afternoon light.

Kadena's farmhouse stood barely inside the gated property. In better times, the building was a brilliant white, but now it was an ashen gray, as if the siding gave up on this world long before its owner did. A wooden fence followed the property line, the occasional timber missing or broken. Kadena's wife passed too early in life, leaving her with a small fortune, an active business, and a massive property to maintain. Opening the gate, Kadena moseyed forward, her apron and work belt slung along her hips. Her raven hair was a whirlwind on her head. "It has been too many days," she called to Theodora's approach.

The words were not said in malice but merely observation. It *had* been too many days. Theodora wanted to return to these fields sooner, but the klaud was too good to pass up. Theodora's travel had been limited to in and around the capital.

Theodora nodded her head in acknowledgement, trying to find the right words to say to Kadena, knowing her wife had passed away around this time many fiedations ago. But instead of providing an appropriate response, she tugged a chain from under her shirt. At the end of the chain was a thin metal cylinder. She brought it to her lips and blew a tune. It barely danced across the field before a horse appeared from a cropping of trees. The mare was an ash color with matching mane and tail. As she raced to them, it was like watching smoke whip through the air. The mare slowed when she approached, swishing her tail in greeting.

"Hey, there." The whisper fell into the breeze between the two of them. Theodora took a moment to rest her head against Down River's coat, the warmth providing a sense of resolve.

Even though Theodora couldn't find words for Kadena now, she had spoken them when it mattered. However, those words had come at a cost, and this awkward silence was the result. Shortly after Kadena had lost her wife, Theodora had been there to help her around the farmhouse on days Kadena struggled to make it out of bed. Not necessarily because they were friends, but rather because Theodora knew the aches of the heart, more than she ever wanted to.

She raised her eyes to meet Kadena's and saw the depth of her sadness. "Are you okay?"

Kadena stayed silent for a moment longer. For eight fiedations, Kadena had borne the title of widow while she continued to force smiles. There wasn't an encounter when Theodora failed to ask this question because time never healed the deepest wounds.

"You are a kind soul, Theodora."

A chuckle escaped. "A kind soul, I am not. But I appreciate the give and take involved in this short window of time we call life."

A smile drew on Kadena's lips in understanding as Down River, who was losing her patience, nudged Theodora's arm and dropped her muzzle to the pouches along her hips. Pulling out a few sugar cubes, Theodora handed them off.

"Well, I can't keep her waiting here. You know how she is." Theodora grabbed a bit of mane and slung her body up. "I won't keep her long."

With a nudge from her boot, the mare jolted forward through the open gate. Theodora focused on her own body to ensure she constantly shifted in rhythm as each hoof pounded into the ground.

They raced across the fields until slowing at the banks of the Undost. The water rushed across the rocks, crisp and clear, a rage of power as it made its way over the land, dividing the continent. Theodora plunged her hands into the cool water, bringing it up to her face as she attempted to wash away the endless thoughts in her head.

She strolled along the banks, Down River following at a leisurely pace behind her. As she reached the top of a knoll, Theodora stopped to take in the view. In these rare moments with Down River, Theodora could escape the mundane and ponder her ideal future. Anywhere else, she was trapped in the hourglass, forced between the sands of time, as gravity pulled them closer and closer to their end.

The blades of green swayed in the teasing breeze as the sky began to bleed pale pinks and yellows. Fiedel began to fall behind the castle and Petram took the mantle from the west.

But as Theodora's body stilled, her mind raced. There was no way to focus on the future forever, as the memories of her past returned.

When she was a child, she would have full days of adventures before returning home carefree. Even when her mother had attempted to reprimand her for the mud and grass covering her body and clothes, Theodora remembered the light in her mother's eyes, the smirk on her face.

Down River snorted nearby, bringing Theodora out of her past. She fitted her hands onto the mare's withers and pulled herself up onto her broad body. Theodora directed the horse forward, but their return to Lume was not nearly as fast as their departure.

Delivered

Maddox was lounging in an overstuffed wing chair an exquisite shade of deep burgundy velvet. Although he'd thrown both his legs over the arm of the chair, he still had a sense of posture. Grace, even. The room was dimly lit, the walls a muted shade of brown. Smoke from pipes and cigars swirled around the patrons and surrounded them in a fog.

Valix and the twins, Gemma and Jemmie, were at one of the game tables. Shot glasses were littered amongst decks of cards, large glasses of beer in front of all those playing.

The refreshing drinks helped to ward off the humidity, even if their beers were watered down, but it failed to bring out any real competition at Ludi Votivi.

It was worrisome that the air was so warm. In the depths of the tunnels, it was rare, not only because of the cool ground around them, but also because of the ventilation system working to help circulate the air. Or at least that was what it should have been doing. Maddox mentally took note to send someone to check on the controller relays of the system later to ensure they were functioning properly.

The den wasn't busy, at least not as busy as usual. Maddox had counted eighteen people, not including the den lord, when they walked in a few hours ago. He could have easily gone to the topside for more interesting ways to make klaud, but Seclus was comfortable. It was as close to home any of them would get for the time being.

"Damn Lumens," Valix huffed as he claimed the seat next to Maddox. Sucking on his cigarette, the glow highlighted the irritation in his frosted eyes.

Maddox's grin widened before he brought the glass of cinnamon spiced liquor to his mouth. When he took a long gulp, the liquid burned as it moved down his throat and warmed his stomach, leaving a numbness on his lips and tongue. It brought a fire to him, burning away burdens.

"We won't have to be here much longer, Valix; don't worry." Maddox lazily drew the words out.

"Worry, sir? I don't know the meaning of the word." Valix chugged another beer, the pale golden liquid escaping past his lips and dripping down his chin.

Laughter broke out from a different table, grabbing the room's attention. A young lad had finally won his first match. The room erupted in applause and cheers.

"The boy looks barely old enough to have weaned off his mother's tits," Valix bellowed as the commotion began to die away, the empty glass swinging from his fingers. The faces of those in the den turned to Maddox, their universal question shown across heavy eyes and tired cheeks.

Maddox raised his glass in a toast. "Another round of drinks, Gustavo." A burst of excitement filled the room as people clambered over to the bar to put in their requests.

These people Maddox had come to know. He built and engineered and worked with the men and women alike in Seclus. This city was in his control, and he was determined to find ways to improve it, refusing to settle for less. Complacency got you killed.

Valix shifted his weight in the chair. Maddox leaned his head, raising an eyebrow in question, the scents of sour beer and cinnamon liquor mixing between them.

"Those women are gawking at you, sir."

Maddox turned his attention to the two women in question, their curls of auburn and blonde entangled as they whispered into each other's ears. There were no cards spread across their table, but there were a few empty glasses, the rims stained a deep red from lip coloring.

The blonde locked her eyes on Maddox, a blush emerging from her cleavage to her neck and face. She bit her bottom lip before taking a swig from her glass. To another man, maybe it would be worth the time, but to Maddox it was another dead end.

Maddox lifted his glass in a toast to the women before draining it to the dregs. He threw the glass at Valix, who deftly caught it with one hand.

"They are both yours, my friend." Maddox offered him a farewell.

"Don't mind if I do," Valix mumbled to himself, jumping up from his seat to take advantage of the opportunity.

"Gems!" Maddox called to the twins who were still considering their hands at a nearby table. Without a second glance, they stoically rose, throwing their cards down before pushing their chairs backwards, the legs scraping ceremoniously across the floorboards.

The twins were tall, nearly reaching Maddox's height, with hundreds of little black braids twisting from their heads, pastel pinks and purples peeking from underneath. Both moved with endearing grace, and their endless silence was most alluring.

"What about Valix?" Jemmie asked as she neared.

"He has to go release some tension, and I'm certainly not going to do it for him."

Gemma laughed faintly as she interjected, "Should we wait a moment? It won't take him long."

The corners of Maddox's mouth turned up before he shook his head. "Come on, Gems."

They stepped out of the den and into the darkened tunnels of the underground city. The air was cooler here than within the den, meaning the ventilation issue resided within the building itself.

While preparing to descend deeper into the belly of the ground, Jemmie flagged Maddox to stop at the sight of a Satelle walking down the tunnel entrance. His golden jacket sparkled from the various chandeliers hanging below the cavern ceiling, and his cape billowed about him with his descent. It was odd to see Satelles in the tunnels, especially at night. The apprehension was apparent on his youthful, freckled face.

"I'm assuming you're Maddox." The Satelle's quivering voice snuck through the air. Sweat beaded along his overbearing forehead as his eyes jerked nervously between Maddox and the twins. Before Maddox could confirm, he continued, "I was told you would have on a red cravat." The Satelle motioned to Maddox's throat as if he needed to prove his observation.

"I am," Maddox responded quietly, glaring downward. Even though the Satelle had the higher ground, he still didn't reach Maddox's height.

"Prince Rajveer sent me to deliver this to you." The Satelle thrusted him an envelope stamped with an unbroken royal seal.

Maddox took it. "You can tell your Prince it was delivered. Dismissed."

The Satelle let out a sigh of relief and turned quickly on his heel. He gripped the hilt of his stelgladio as he moved back up the tunnel to the outside world.

Maddox looked around to ensure they had no company. He ripped the green wax of the crowned seal, pulling the paper apart to read it. He quickly took in the prince's scrawl of words, a prospect contained in the swirl of ink on parchment. Maddox's lips twitched.

As he lifted his gaze upwards through the tunnel, Petram stared at him, but Maddox didn't look there tonight. Instead, he looked beyond, as if he could see all the way to the castle where the spoiled drunk sat.

Silence

Rajveer kicked at loose stones in the early gray light as he walked patrols with a handful of Satelles. The slight chill of the air attempted to pierce through his jacket as a thick fog floated around them.

He didn't need to walk with them. It was unheard of until he had started fiedations ago. After his would-be bride betrayed him the night before his wedding, he had spent a petrik locked away in his room. His mother persuaded him out of the castle and Amicus had invited him to join the Satelles on their patrols. It quickly became routine for him and one he wasn't willing to let up.

What had started as following his mother's simple request turned into a realization that the patrols got him to the people. It gave him the opportunity to talk to his citizens about their sources of distress and possibly lend solutions. Mostly, the act of listening was enough. For now, it was all he was capable of, at least until he gained the power to make significant changes for them.

A few of the Lumens would offer disinterested glances, likely out of fear or respect for the king. Some, and luckily most for Rajveer, were giddy as schoolchildren who rambled so quickly, he was unable to voice his thoughts through their thousands of questions. But again, he didn't have to. He wanted the people to know he was here and with them. He saw their discomfort and they were not alone.

Rajveer and the Satelles circled around the northern side of the capital, walking past the Reme Temple, when a cloaked individual stepped out to lock the door. When they turned around, it was not the senior official he was expecting but instead, a young woman he had never seen before. Curiosity got the best of him and Rajveer called out as she began to turn down the street.

"Excuse me, miss."

The woman turned back to face them and immediately dropped to a curtsy. "Sir. How can I be of service?"

"Where is Noble Zyair?"

"Oh, he came down with a terrible cough a few days ago. But not to worry, he has been drinking redwood tea and is on the mend. I would expect to see him back here in a few days. Was there... is there something you need? I might be able to help."

Her hands were a lively accompaniment when she spoke, forcing the hood of her cloak to slip off her head. Rajveer immediately flashed a grimace at the sight of her hair.

"It's jarring, isn't it?" She continued shamefully, as she attempted to replace the hood back into position. Her hair was a mixture of whites and shades of gray, while her face made her appear younger than Rajveer. It was a battle between old crone and young beauty, and he wasn't sure which was winning.

"I mean... sure, at first, but it's the contrast that makes it appear that way. I think it's uniquely beautiful."

Rajveer noticed a peak of a smile before she glanced down at her interlocked fingers. "If there is nothing I can help you with, I guess I should be going."

"Of course. Always a pleasure, miss." He placed a hand over his chest with a nod.

He and the Satelles continued around the bend in the street, their patrol taking one of the paths leading into the first row of residences. The building lights slept at the hint of morning. Pots of wilting flowers decorated porches and front steps, fading away from the heat of the previous days. In a short time, Lumens would replace them with dahlias and chrysanthemums. The jewels in the stone walls would reflect the warm colors of the petals, making the whole capital glow in astrum reds, oranges, and yellows.

Rajveer was so focused on the decorations adorning the buildings, the wreaths of gold and green or the occasional drunken gnome statues placed next to potted shrubs, he hadn't realized the path they were taking until it was too late. He stopped short of one of the alleys, the memories of eight fiedations ago crawling into his mind like the fog across the streets, filling the crevices it could reach.

∴ ∵ ∴

The heat of Fiedel made him uncomfortable, the setting light blinding him with rays of gold as they stretched beyond the capital rooftops. He glanced down at the glimmer from the cobblestones, a single klaud wedged between two stones. The small pebbles scraped at his bare knuckles as he pried it free.

"Think you're going to win yourself a date with that?"
Amicus' voice teased. Happiness settled within the words as he walked next to Rajveer.

"Ha! The Lumens pay me."
"Yeah, for disappointment."

Rajveer's laugh filled him, starting from deep within his chest. Amicus, and the Satelles who flanked them, joined in as well. The laughter flowed out of them like the Undost River. Although Amicus was technically a member of his father's guard, he had always been like an older brother to Rajveer. Together they'd engaged in bouts of wrestling in the library, races in the halls, lessons on how to wield a stelgladio, all things any older brother would have taught Rajveer.

The royal company condensed into the alley behind them. Rajveer stole a glance in search of the horse bearing his father. As he turned back around, a woman working on the front steps of a home caught his attention. Her face was angled down so he only saw her silhouette, but it remained unobscured, as her hair was hidden beneath a headwrap. She rose to her full height, turning in his direction and forcing Rajveer to steady himself as his feet attempted to stall beneath him. Fates, she was exquisite. Her skin was darker than his own, reminding him of toffee, and for a moment, he wanted to know if she tasted the same.

She was decorated in an elaborately colored dress, bright fuchsia and warm greens spattered with gold lace. The definition of her curves was lost in the heavy fabric, and Rajveer craved to discover them with his touch. Pushing past the Satelles who continued forward, Rajveer approached the woman, stopping at the bottom of the steps.

"Miss…" The words fell heavily from his mouth, barely audible over the wild thumping within his chest.

Her body turned closer to his and he lifted his face to hers. A strand of her dark hair threatened to escape past the fabric, and Rajveer chastised himself when his hand twitched to push it away.

"Sir? Oh…" Her voice tripped down the stairs. "My prince." She immediately placed her hand across her chest, bowing. As she raised her head, Rajveer caught the smell of rose and a spice he couldn't identify. It pulled him forward like a magnet, a tether.

And then she looked at him, a small smile on her face, and he lost all sense of thought. It wasn't as if time slowed – no, that felt too cliché.

"I feel like I'm at a disadvantage." Rajveer forced himself to grasp words. "You already know my name, but I *don't know yours." He combed his fingers through his hair, tangling his locks as badly as his nerves.*

Her smile faltered as the king's horses clomped closer.

"Alouette." Her voice was soft, timid, and Rajveer had to work to hear the name. The letters wrapped around his heart as if they were making a sacred vow.

∴ ∵ ∴

"Sir?" One of the Satelles dragged him from the vision of memories. Sweat scattered along his hairline, and Rajveer pulled a handkerchief from his pocket to dab it away. But when his metallic hand touched his skin, it jolted him further. "Are you okay, sir?"

"I'll be fine." Rajveer let the words fall quickly, but they were reactionary. He didn't feel fine. The memories weighed heavily in his mind.

"Do you need to head back to the castle? We can have the horses readied."

"Trust me, I'll be fi–," but Rajveer cut off with the sound of boots running in their direction. The Satelle patrol moved their hands protectively to their hilts and waited before another Satelle ran around the corner. The Satelles were about to relax until they noticed the evident tension on the face headed for them. "There is an attack in one of the fields," the approaching man said. "It's all burning."

Rajveer and the Satelles spurred into action, clamoring into the saddles of their horses. The hooves of the royal company were a symphony as they followed the streets, the sound echoing off the high buildings until they broke out into the main square. The occasional Lumen ducked to the side, fear mirrored on their faces. The capes surrounding them whipped back and forth, shadows following in their wake.

Rajveer whipped the reins, pushing his steed harder when Amicus barreled in their direction on a horse of his own. The patrol's horses reared at the chaos and commotion.

"Stand down," Amicus called to the Satelles as Rajveer tried to tame his own horse.

Adrenaline coursed quickly through Rajveer's body, removing his ability to speak. His steed continued to move erratically, feeling his energy.

"I'm sorry." The words finally made it to Rajveer's mouth. "I didn't hear you properly. You said, stand down? As in, don't head to the fields where someone has set it ablaze?"

"That's correct. King's orders."

Those last two words were filled with as much authority as Amicus could muster.

The unease of the other Satelles was palpable.

Rajveer wanted to yell at his Satelles. He wanted to scream at his father. He wanted to roar at the fates, but he forced it away, swallowing hard against the disputes forming in his throat.

In silence, he turned his horse in the direction of the Digere, craving nothing more than a drink to chase down the unsaid words while he prepared to end this reign.

Task

Theodora untangled herself from her sheets, releasing her fingers from their tight clench over her dagger. Small indentations from the gems and woven metal marked her palms.

She stretched her body out, the springs of the mattress popping with the movement. A quick breeze pushed aside the small cut of fabric tacked above the window. The pale gold filled the room with not only light but also heat.

It might have been peaceful, except the hungry baby below had moved from soft whimpers to full blown wails. The children on the floor had started their races around the small space, and she was certain the oldest was still winning.

Her bare feet scraped along the wooden floorboards as she walked through her home, headed for the small washroom. It was small, but it was all hers. Well, except for most of the furniture, which had already been there when she had moved in: a table and chair, a bed, and a small collection of cabinets and countertops. Her only additions were the papers neatly arranged on the table with notes, a handful of books in the corner untouched over the last few petriks, and a lush wing chair.

And the array of scattered weapons and clothes.

Cleaned and dressed, she returned to the window to push the curtains aside, unlocking the window mechanism to allow the single pane glass to open it fully. Reaching out to the exterior window trim, she pulled the shinegun from a holster she had attached there. The discs on the top were a shiny obsidian, having charged within the few hours of morning fiedelight. She bent down, slipping the shinegun into her boot, activating the mechanism from within. She reached to the nearby table, pulling her dumgun, with its long, thin metal barrel tinted blue and a dazzling filigree with a grip of pale oak, and slipping the mechanism into her right boot.

She added her gear like armor. It was a ritual: the strapping of waist pouches, tightening of thigh bands, slipping of hot spots and flash crystals into leather, and sliding of her dagger into its sheath. Wearing every piece was the only way to face the capital. Being comfortable was what had cost her parents their lives.

The door whined as she exited her room and entered the hallway. The weight of the world mounted her shoulders as she descended the stairs. Before she had moved in, the plaster had probably once been bright, but now layers of grime had slowly seeped in. The stairwell lights blazed with life, accenting the chipped walls, the missing nails of the floorboards, and the holes scattered throughout both the floor and walls.

She left the building, squinting in the bright rising Fiedel as she searched the street before her. She resided at the end of the building row, though there was no difference from the countless other buildings around her except for the number on the outside.

The noise of the vendor carts headed from the square were loud, echoing in her ears. She followed the stream of people from the residential area on the outskirts of the capital.

The city was shaped in a wobbled circle with the grand fountain in the center of it all. Immediately surrounding it were taverns and shops, the clergy, and medicinal makers. Surrounding those, on the outskirts, were the residences, the houses of Lume protecting the king's prized possessions.

Theodora continued the path of cobbled streets and lampposts, the fiedelight reflecting on the sparkled jewels in the building faces. During her youth, Theodora had found the early morning walks to be the prettiest. Of course, she had spent far more time in the shadows and away from the lurking eyes of the Satelles.

A lean Satelle of pure height shouted at her across the street. "You, Theodora!"

She halted and forced a smile on her face. "Yes, sir?"

"Where are you headed?"

He approached, and Theodora noticed his hair was set in the neatest of rows lining his head.

"I'm headed out for morning coffee."

He squinted at her for a moment, attempting to find a reason to reprimand her, no doubt. However, finding none visibly apparent, he dismissed her with a terse nod.

She had debated giving him a terrible rendition of her salute, but settled for a mock bow of disrespect.

For a world with no uprisings or threats, at least not until Seclus erupted, the royalty had a history of recruiting large amounts of Satelles. These Satelles were stationed across the land, keeping tabs on the cities, maintaining compliance. But a decent number stayed at the capital, earning a prestigious position on the king's guard. And a cute cape, although with the advancements of Seclusian's weapons, she imagined all the Satelles wore them now.

She arrived at the row of local taverns, choosing her usual, the Digere. It was crammed into its spot, and some of the stones on its face were shattered or falling apart. Even the sign above the door appeared attached only out of sheer stubbornness.

The food at Digere was manageable. It wasn't outlandish, and the price was decent. But Theodora was comfortable here. The patrons were a little more reckless and the Satelles who did show up would usually turn an eye from the Lawless within.

She entered the dim tavern, with only half the lamps lit and limited windows. Long shadows grew along golden floors, making them appear an inky black. The entire perimeter of the room, save for the bar in the middle of the back wall, was lined with table booths with wooden benches on either side and towering backs. The wood jutted from the benches and climbed halfway up the wall, providing a backdrop. There were a handful of tables set up in the middle of the room, some with two chairs, others with four. The little space of bare wall left above the booths was covered in art: handwriting, doodles, elaborate murals, everyone attempting to leave their mark behind in this small space of the world.

Men and women alike sat on tables, leaned against booths, or knelt on chairs. One booth somehow held ten bodies as they tiredly picked at their food, half-filled mugs littering whatever spare space they had left.

Theodora locked eyes with the tavern keeper behind the bar, and he gave her a brief nod. Julian Albani was a stout man with a balding head, although he hadn't realized it yet. The hair on his arms, visible from his rolled-up sleeves, was thick, matching the grease caking his apron. Behind the bar top, which had no stools before it, were shelves that reached the ceiling. Each shelf was filled with unopened bottles of wine and other spirits, the dust having made the labels illegible. The only drink served, other than the morning orders of coffee and water, was diluted beer from a keg, stamped boldly with the king's crest.

With her boots sticking to old spills and new ones, she worked her way over to her table. The table Albani held for her had a burnt-out bulb above it. It was shoved behind a post, askew due to the wall, and one chair had been thrown with it.

Although its location appeared awkward, it allowed a wide view of the room. The table was a long-standing agreement they had. Albani gave her the table, and she ran errands for him. Her eyes moved around the crowds of suits and trousers, full skirts, and waist belts. She picked up quick exchanges. A word here, a word there. She made note of the few in the room with blades tucked barely out of view, a handle hidden by the fabric of a jacket along the waist or a hilt barely above the top of a boot. Although brave souls ventured here, they were nothing compared to those in Seclus.

A waitress delivered a hot mug of coffee to Theodora. "Can I get you anything else at the moment, miss?"

"No, thank you," Theodora replied, lifting the mug to her face, the strong nutty aroma assaulting her nose. The waitress gave a brief smile before weaving away.

Theodora scanned the Lumens again when her gaze caught on a Satelle entering the tavern. Her golden jacket glowed brightly, commanding the space. The conversations and laughter dropped low, accompanied by worried glances as the patrons picked at their food. Theodora rolled her eyes as a groan escaped.

Danika wasn't the most beautiful woman she had ever seen, but she wasn't terrible-looking either. It was as if she was born for the role of Satelle. Her respectable height and average build were accented with waves of golden hair on one half of her head, brushing past her shoulders. The other half of her head was shorn close. Her eyes shone as brightly as her jacket, a brilliant shade of green.

Gold and green, the colors of Lume.

Two petriks ago, they had courted each other, or she assumed that's what Danika had called it. Theodora had toed the line of respectable Lumen citizen and Seclusian Lawless. It was bad enough she embraced the Seclus life, but to also be a Lawless? For the Lumens, it was unheard of.

Theodora used their relationship to her advantage to gather intel for tasks. Danika had seen her as something worth fixing. To Danika, Theodora was a Lumen who had merely lost faith in the crown, and maybe even the fates. Danika offered Theodora a new life, a fresh start. All Theodora had needed to do was sacrifice her name and freedom.

Theodora had been smart enough to not get caught, at least not with any severe crime the Satelles found worthy enough to punish. Those who did met the consequences, usually involving some type of body mutilation, or, if you were lucky, death. Danika had wanted to prevent Theodora from earning such a reward.

"Theo!" Danika bellowed over the crowd. Whisking the rich forest green cape away from her body, she sat on the tabletop, placing a hand on Theodora's knee. She didn't attempt to hide her glance down Theodora's body before speaking again. "It's been a while. How are you? Still visiting Seclus?"

Theodora reached forward, taking Danika's hand off her knee and throwing it aside. "You know I can't stand when people call me Theo." Taking a moment to let the warning settle between them, Theodora crossed her arms over her chest. "I'm fine. And Seclus is the same as always."

"I was told you weren't visiting as often. I thought maybe you were finally switching to the topside. Taking my offer seriously." Danika's eyes sparkled with the possibility.

"Thanks, Danika, but no. I've told you I have no desire to be a topside partner. Now, I would like to enjoy my coffee. I am sure there are other people here who would love a visit from Satelles."

"Are you still—you know?"

"What? No, I don't know."

"Are you still doing jobs for Seclus? I know there is a new one no one has taken up yet. Word from Seclusians is the task is being held for you."

"You know I'm not going to answer that. You're dressed in the king's garb. I won't acknowledge anything."

"What about without the garb? Come on, Theo! You always talked to me before."

"No, you made assumptions based on the small details I gave you."

"But I want—"

"Danika, please!" Theodora lowered her voice as she noticed the curious eyes looking in their direction. "Please, stop." It was barely a whisper. "You may be able to use your power to get others to do your bidding, but not me. Don't you have some patrols or something far better to do?"

Danika only shrugged, removing herself from the table to mingle with others in the tavern as Theodora sipped from her mug, attempting to determine what Hakon's task could possibly be.

Calculations

Maddox slipped his arms into his charcoal vest, pulling the fabric closed over his muscled chest as he moved to the three silver buttons. He shifted to face the full-length mirror, which stood within the corner of his room located within Previt, a building of residence apartments located in Seclus.

As he fastened the top button of his shirt, he tugged the crimson bolo tie from the corner of the mirror, wrapping it around his neck. Tightening it in place, his glance caught the reflection of the closet behind him. A pair of crooked trousers seized his attention, and he immediately turned to fix them before the tie was secured around his neck. After correcting the fabric on the hanger, he swiped at a speck of dust. Clean, meticulous, with an air of being unlived in: it was exactly how Maddox preferred his home to be presented.

Although this was nothing compared to home.

Maddox folded his shirt collar down before sliding his pocket watch into his vest pocket. Catching the time with the movement, he knew his companions would be wrapping up their breakfast together in the first-floor tavern attached to the building.

He descended the stairs, the lampposts shining brightly into the wooden stairway. He counted forty-eight steps to the bottom, adjusting his gait for step twelve where he knew the wood squealed. The dull mumbling of the other tavern dwellers intensified. He followed the scent of the coffee, counting the faces of those in the tavern. One cook, two waitresses, twelve men, and the twins. He took note of their placement, confirming the exits, before stopping in front of his companions' table.

Valix lounged in the chair, an ankle rested on his knee. "Sir."

"Well, you're here, so I guess the night wasn't *that* remarkable."

"The night was fine. I don't know what you're talking about," he mumbled into his mug before downing the last of his coffee.

"Don't let him fool you, Maddox. He was back home shortly after you turned in," Jemmie interjected.

"Traitors." Valix set the cup down definitively before rising and turning to Maddox. "Did you hear from the Satelles?"

"We received word to meet with them at Digere. They have secured a space on one of the upper floors."

"An appropriate middle ground. There is no way the Satelles would allow Rajveer to come to Seclus."

"And no way they would allow us to come to the castle," Gemma added.

"Do you blame them?" Maddox asked the group as they exited the tavern. But the twins fell into their silence, trailing behind the two men.

Valix opened the door leading to the underground city before continuing the conversation, "No. I am curious to see what Rajveer will have to say though. It will be fun to speak directly to their prince for once."

"What, tired of writing love letters?" Jemmie could barely finish the question before she snickered.

"It's the only pursuit he gets," Gemma continued. Her laugh was drowned out by a woosh when the ventilation system kicked on as they strode out of the building.

The system was a necessity for the survival of Seclus. Constantly breathing in the dust and dirt would be unacceptable, and a solution was one of the first things Maddox had designed. After the ventilation was created, they had moved to the structure of the underground tunnels: the buildings carved into the walls of the caverns, the tiers connecting the two, the lighting, the plumbing; all physical things crafted and built both efficiently and methodically. The politics and current events were what Valix was there for.

Maddox and Valix were reared together, both brought to Seclus initially by Valix's father, Stavros. They were expected to learn about each other so deeply, there would be no secret left unshared between them. They had wandered the hollowed-out halls, explored the darkened caves, climbed the stalagmites, and waded in a small pool gently illuminated by the creatures within.

When their childhood was shed for adulthood, Valix's father left them, and in his place, the twins joined them at the undersigned city. They spent less time with Seclus' hidden treasures and more time building, fixing, and adapting.

"Has your father said anything recently?" Maddox inquired as they slipped between people headed in the opposite direction, but any response from Valix was cut short from a Seclusion.

"Sirs?" she interrupted as she bowed, strands of auburn falling loosely from her tight ponytail. "Sorry for the intrusion, but we have an issue with one of the pulleys. It's acting up again?"

"The same as before?" Maddox responded.

"Yes, sir."

"I told Sanjay he can't tighten the main mounting bolt too much. You will need to loosen the bolt and possibly the tension screw on the side as well."

"I'll be sure to let him know. Thank you, sir."

Maddox nodded in her direction before she slipped away. He turned back to Valix, returning to his previous inquisition.

"Stavros. Anything?"

"He is getting impatient with Lume…"

"How? They have known this was an investment. If they want the resources, they will need to become patient."

Valix showed his palms in mock defense. "The committees aren't in agreement with him. I'm letting you know my father is growing impatient. He is trying to garner support for an injunction."

"Your father is a piece of shit."

"Tell me something I don't already know."

The cavern was filled with paths and bridges connecting to various buildings, a complex map for anyone new who ventured down here. A city of grandeur dropped under a rock required use of all the space manageable. But although it was underground, it was clean. Smooth stone walls and immaculate paths. The scent of the shops mingled with the stagnant air until the ventilation kicked on again, and with a whoosh, swirled it away, leaving the scent of freshness and sometimes rain depending on the weather.

The four of them walked through a group of Seclusians, all giving slight bobs of their heads or tips of the caps in acknowledgement. Compared to those who lived topside, the Seclusians were machine-like. They shared similar clothing choices, trousers, and button shirts. Suspenders and work aprons for those actively working, vests and jackets for those who were on their way to business. There was practicality in every decision, and every one was made for the betterment of the city. There was no competition among the people, only unison.

Breaking away from the slumber of dwellings, the four descended the stairs of an upper tier into the shops. Above the seeping of air, the sound of workers rose and echoed off the walls. There was the clanking of metals, the grinding of gears, and the bubbling of liquids contained within vials. Metal and machines worked in harmony at their hands in the forging of new ideas and possible solutions.

Maddox and his companions headed to the tunnel entrance, the square at the bottom a close replica of its capital brethren. But no matter how many times he passed through this one, Maddox couldn't stop himself from glancing up at the chandelier hanging from the ceiling.

It was his pride. While Valix helped with the businesses in the square, Maddox spent his time at a table sketching, planning, and detailing every curve and line. until his forearms and hands coated with charcoal as calculations were made. But alas, he had found a way to use the creatures from their pools, to capture their emitting light permanently within glass globes.

Making it to the topside, they passed the rocks jutting around the entrance. Maddox looked at each of his companions in turn as he pulled his pocket watch from within his jacket. "Let's go see what this prince has to offer."

Another

Rajveer paced in one of the storage rooms above the Digere, the sound of his tapping fingertips intensifying as time passed. The Lawless attack from this morning and his father's brash decision fueled his irritation.

Amicus had raced off to ensure the perimeter patrol of Lume was solid, as well as to secure the castle. Rajveer took a mental note, again, to remind his father of the ludicrous decision to have the castle so far from the capital, but knew it ultimately would fall on deaf ears.

With his Satelle Captain fulfilling his required duties, it left Rajveer with his three Satelles: Danika, Jude, and Miles. The night of their Satelle ceremony, after taking an oath to the crown, they had sworn their allegiance to Rajveer as their future instead. They stood in the room with him now, the peeling cream walls making Rajveer claustrophobic by the sheer amount of gold from his Satelles' jackets it reflected in the mid-morning light peeking through the grimy windows.

A double knock sounded, and the space quieted. Rajveer's pacing ceased, and he shoved his metallic hand into his pocket as they all turned in the direction of the door. Danika brushed her cape away from her hip, gripping the hilt of her stelgladio. She looked back, holding the stares of Miles and Jude, who stood protectively in front of Rajveer, before she opened the door.

Rajveer nearly flinched at the sight of the four people now filling the threshold. They had a unique look, and Rajveer couldn't deny their attractiveness. The man in front had thick waves of hair, but it was his icy eyes that caught his attention.

He stood with both hands deep in his black trousers as if he didn't need easy access to the shinegun strapped to his hip.

Two women who were mirrors of each other filled the sides of the frame. Their dark skin was a stark contrast to their male counterparts. Their braids fell past their shoulders over their matching black leather jackets. White cotton shirts peaked out from where their arms crossed across their chests.

And behind all three was the one Rajveer assumed was Maddox. Although he had never officially been introduced to him, Maddox's reputation preceded him, including his dark as night hair and colorless eyes. Although their eyes connected, Rajveer's metal hand, the one crafted in Seclus, burned as if Maddox's gaze was there instead. Rajveer assumed Maddox was the one who'd overseen its design. But as with everything relating to Seclus, it was all assumptions, false stories, and fleeting gossip.

Only the two men walked inside, forcing Danika away from the door and deeper into the room as they shut themselves in. Rajveer was curious why they had brought the twins only to abandon them there, but the thought only remained in his head. For a moment, nothing sounded in the room except the faint musings of the tavern below.

Miles was the first to crack the silence. His youth showed on his face, the eagerness bright in his round eyes. "Uh, my Lor-Lord Maddox—"

A devious smile appeared on Valix's face as Danika silenced him with a hard glare.

"What shall we call you then?" Rajveer interjected.

"Maddox and Valix will do. Don't worry about the twins out there. Their importance is elsewhere; they won't be contributing to the conversation."

"Or sir," Valix continued for Maddox, the smile failing to fall from his face. "How may we be of service, Prince of Lume?"

Rajveer stepped forward. "We would like to hire you for your services."

Silence filled the space, like a fat cat that had found a sun spot. It took its time sprawling into the room.

"Which of our services would that be?" Valix responded, taking the lead in dragging the answers from Rajveer.

"Can you swear to secrecy?"

"Listen, kid, this isn't the first time we have been asked to participate in questionable scenarios. Do you want our help or not?"

"You will address him as prince." Danika was quick to intercept, yet Valix remained quiet, his inquisitive eyes narrowing at Rajveer.

Rajveer had to force himself not to twitch in any capacity, but with his nerves building, he felt erratic. He wanted to pace, to tap his fingers, to comb his hand through his hair, anything other than standing still and staring down some Lawless.

"We want you to assassinate the King."

"How original," Valix mused.

"Look, we have our reasons, and if you want the klaud offered then you will do it without question—"

"Without question? You do realize you're the one at risk here."

"At risk? You four are Lawless. You roam our streets and threaten our people. I could have you detained on—"

"Detain us? I would like to see you try. There is a reason your father, your king—"

"Do not speak of the king in any sort of tone!'

"Me? You want to have him murdered!"

Rajveer reined himself back. He tried to reel in the emotions—but isn't that what this was? Killing his father was pure emotion.

When he'd made the decision last night, he was in no state to do so, but when morning came, his bed empty from the woman who took her payment and escaped during the night, he felt relieved. He was no longer anxious for the future. Rather, now he was excited for the possibilities it might hold. He had bigger ideas for Lume; he just needed the throne.

Rajveer, exasperated, continued. "Look. We have a common interest here. We know you want to disrupt the power. Show me your worth, your loyalty at least to this request, and maybe there is potential for a future partnership."

"Explain how this benefits us?" Maddox asked.

Without waiting for a response from Rajveer, Valix continued the interrogation. "Have you any thoughts as to how? Why seek out Lawless anyway?"

A thoughtful quiet filled the room. Although Rajveer had brought the task before his Satelles, they hadn't gotten to a plan for the actual assassination. Fates above, they hadn't even thought of how they would persuade the Lawless.

Amicus had attempted to bring it up earlier, but Rajveer had dismissed him, saying he would figure it out when the time came. Now the time was here, and he could feel the stare from his Satelles. He had to show them he could be king. Clearly, he'd proved something to them before and he couldn't fail them now.

Danika began to speak, but Rajveer waved at her to stop.

"We are willing to offer a large sum of klaud. Ten thousand pieces, to be exact. We will also wipe away your past transgressions against the crown, but it will only be a clean slate to start. Although you have skirted my father's punishment in the past, it ends with my rule. Any new acts will have consequences."

Rajveer met the stare of the two Lawless before him, hiding the fact that he was nervous.

"We're gonna need more than that," Valix began. "It's a large amount of wealth, but we're gainfully employed. Also, why a clean slate with the threat of future punishments, if your father lets us continue as we are?"

There was the crux of the situation. He needed the Lawless involved, as he couldn't ask one of his Satelles. It was risky enough allowing more than one person to know of such a plan. However, Rajveer knew he wouldn't be able to kill his father either, regardless of the man's endless fiedations of abandonment not only of himself as a son, but also of the Lumen people.

The Lawless, especially these four, had a history of insinuating extreme reactions in public. Rajveer could count on one hand Valix's public outbursts alone within the past solta. Some were less egregious, like attempting to make a flamethrower from his mouth filled with liquor, while others were downright despicable, such as the strong evidence that he had killed a family in their sleep. Yet when he presented any of these crimes to the king, punishment never dropped. Instead, the king had dismissed the charges without explanation.

Rajveer saw how these people continued to hold onto their power and knew it was only a bargaining piece. "What if I let you help with Seclus? With me taking the throne, Seclus will need to be addressed. You can be on a council for discussions on how to proceed, whether it be removal or establishing it officially as a city."

Valix turned to Maddox. Rajveer had clearly struck something they sought. Maddox rubbed his chin and returned a brief nod.

"We're in agreement then." Valix paused for a moment before continuing. "And the plan?"

Relief briefly washed over Rajveer before Valix's question registered in his mind.

"Well." It was Rajveer's turn to pause. But unlike Valix who did it for dramatic effect, he didn't have words. "We assumed you would be taking care of those arrangements."

Valix continued immediately, as if it didn't require much thought. "We could drag him by horse from the castle to the capital square. Or castrate him for all the Lumens to witness."

Rajveer was stunned speechless. Was Valix serious? It appeared the Satelles were shocked as well, only responding with widened eyes.

"No? Don't like that one?" Valix went on, coming up with more and more gruesome choices. Rajveer shouldn't have been surprised, though. Valix was merely describing all the punishments his father had ordered against the Lawless, all caught except these four.

"What if we drag him to the square and cut off his hand?"

"No!" Rajveer's own voice rang in his ears before he realized he'd spoken aloud. The word echoed about the quiet room and within his head.

Maddox broke the building tension. "We bring in another."

Rajveer didn't fail to notice the jerk of Danika's attention. No one responded to Maddox; only stared at him to continue with the explanation they assumed was coming.

"We need a way to keep out any possibility of Rajveer being behind this, but we also need a way in. I think there is another, a Lumen instead of Seclusian, who could do it. We find a way to allow them to meet in public, giving the impression Rajveer is interested in her. She can be invited to one of your evenings at the castle, and meet dear old dad, as it were. Nothing out of the ordinary."

"I don't understand. What does a Lumen have to do with this?" Rajveer asked as he glanced at the other Satelles, in hopes for clarification. He had been wary of bringing in his own Satelles, let alone *four* Lawless, and now they wanted to bring in another. His request for assassination was going to be discovered quickly at this rate.

"It is going to be suspicious if you start befriending the Lawless, Rajveer. I know of one who is from Lume instead of Seclus. To the outside, it could be seen as extremely political, or they will think it's another guest you feed, woo, and abandon. But this act will get her into the castle."

Rajveer let the question cautiously escape as his mind raced. "How do we get her involved?"

"Hakon. He already has a request out for a job. I'll intercept him to make sure he has a new task for her."

"How do we know someone else won't take it?"

"Don't worry about it. I'll make sure it's taken care of."

"How will we know it's done, that she has accepted it?"

"She'll find us."

∴ ∵ ∴

Rajveer and the Satelles left the tavern, walking along the cobblestone streets, capes billowing about their muscled bodies as they made their way to the local stable.

The people in the street moved aside for them, stretching their necks to catch a glimpse of the prince. It bothered him more than it should, the way his people felt so defeated and surprised when he showed face, even though he never stopped with his mother's death. Maybe it was why his people viewed him with such idolization and hope.

He dipped his head further into his jacket, grateful for the collar. He turned to Danika, as she had appeared to be the only one who had followed Maddox's plan. "Alright. So, who is she?"

Danika opened her mouth to speak and shut it. Her eyes darted across the street, seemingly absorbing every detail of those they passed. She let out a ragged breath, as if it would help her find the words.

"I'm only assuming, sir. I don't know for sure." She paused again before continuing. "I'm fairly certain, because I don't know who else it could be."

She brought her hand to her chin, and Rajveer could see her internal struggle. The chatter of passersby filled his ears, and the smell of warm bread lingered in the market.

Jude, who was walking in front, turned back. "Is it the one you courted?

Rajveer looked at Danika, but he didn't miss the flash of emotion. Was it embarrassment? Or something else?

Miles began, "Oh, yeah. She was a Lawless, wasn't she?"

"Yes." Danika became defensive. "But she has openly admitted her allegiance to Lume."

Rajveer interrupted, "Is she working with anyone? Does she work for Maddox?"

Miles hesitated a moment. "I don't know. I don't think so…"

Danika remained silent as they approached the stable, quickly mounting her steed and taking the reins. Rajveer threw his body up into his own saddle, pulling his horse near Danika.

"Do you admit you have a fairly damn good assumption of who it might be?"

"Yes, sir," she murmured.

"And you courted her?"

"Yes, sir."

"A Lawless?"

"Yes, sir."

"Then you and I have much to discuss, Satelle."

Seclus

The most difficult part of the journey to Seclus was time. By foot, it could take the entire day to travel to and from the city, depending on where the destination was and the purpose there. Theodora rode hard on Down River, pushing the mare north out of the capital boundaries and across the Undost River. The number of buildings dwindled, and the few remaining were caked in dirt, broken and splintered by the forgetfulness of the crown. The cobblestones had long since crumbled, leaving only a well-worn path of dirt. Even the trees had no desire to be a part of the journey, disappearing with each bounding gallop and leaving flat, endless fields of dying lands in a burnt yellow that stretched out like a desert. A traveler's only companions were the barren ground and rocks.

Thin clouds weaved in the pale blue of the midmorning sky, offering little reprieve from the blasting fiedelight. Theodora wiped the back of her hand along her forehead, removing the drops before giving encouragement to her mare, the wildness of the beast still eager to break free.

Theodora tried to imagine the changes she would witness on her arrival to Seclus. With each trip, she was rewarded with seeing a new advancement their people had created, as they were constantly adjusting their technology for efficiency and ease of life. With her recent hiatus over the last petrik, she was excited to see what would have changed this time.

The barely audible hum of insects and the repetitive thumping of hooves let her mind drift to the first time she had walked into Seclus. A time when she had moved from being called girl or child to miss.

She had ridden to the city with her parents, all three of them giddy with anticipation to witness the new weapon, one which had weaved into the stories of the capital gossip. Upon their arrival, she was tugged deep into the caverns, pulled along too quickly to notice all the world which lay around her. She entered a shop with her parents, and they were shown a prototype of what would be known as the shinegun. It was like the guns they already had, but more compact and with a shorter barrel. Along its top were metallic discs which captured energy from fiedelight.

With the new creation, the gun, common and well-known, had become obsolete. Gone was the need to rely on ammunition and the ability to find or create it, as now they could use an abundant resource, Fiedel. The old weapon quickly earned the slang term of *dumb gun*, and ultimately became dumgun. But shineguns were rare within the capital city, and even rarer across the other cities, given the necessity to have the owner's fingerprints programmed to allow the safety to be released. It meant only those willing to travel to Seclus and register themselves were able to carry. Although Theodora had registered herself to carry and use a shinegun, she continued to keep a dumgun as well. There was something reliable about old technology.

The king, in a mock attempt to show logic, tried to respond to the growing number of shineguns in Seclus with the stelgladios and nocturne capes for the Satelles. Theodora had still been unable to figure out how they learned to create them, as they appeared to be like Seclus' style of weapons. Unfortunately, with the limited equipment only belonging to the Satelles, it left most of Lume's people defenseless and relying solely on the king to protect them. And with the growing number of Lawless attacks recently, this fact made Theodora nervous.

The tunnel entrance came into view, accented with large, cumbersome boulders, obnoxious against the endless flat terrain. Due to the proximity, some Lumens from the capital would travel for a trinket or have an item fixed; and on rare occasions, a traveler from Freta or Vindem would make the journey, but they were typically here only for stories rather than actual goods.

Theodora left Down River near the rocks, allowing her horse to wander the desolate field. The spherical entrance yawned wide, and Theodora looked above her as she passed under the thick metallic opening cut smoothly into the rocks. With each step she descended further into Seclus, the darkened shadows forced her senses to focus. Her gaze shifted quickly along the walls and ground, her ears prickling to every noise that echoed back to her. Here there was no crown or fates. A person survived by the substance of their weapons and sheer luck. Theodora's reputation had been built by herself alone. Also, by the fact that she was Lumen—an oddity she used for every additional klaud it was worth.

The initial tunnel was high, falling open into a huge cave. At the bottom was a replica of the capital market. The shop buildings and taverns were erected in a haphazard circle. Smooth stones of obsidian barely reflected the lights along the lampposts. But where the attention of the topside market square fell onto the fountain, here it gravitated upwards to the ceiling above. From it hung wrappings of black metal fashioned into intricate details racing the width of the cave. Glass globes dripped down, filled with a glowing light of pale blue—the glow, she later learned, was from wormlike creatures found within some of the smaller caverns' pools.

Seclus' expertise in technology was on full display as Theodora passed between the buildings. Doors opened on their own with the approach of an individual, the soft woosh filling the air as people entered and exited. Vendors on the street tapped screens strapped to their forearms to accept payments. Lamps lined the walkways, fighting back the shadows. Drifting from the square deeper into the cave were bridges and tubes reaching for other tunnels. It was a labyrinth of metal and wood, full of the beeps of new tech and the groans of old.

It had taken her time to learn the different routes to get to the places she needed. Although she had visited here routinely for half her life, she knew she hadn't explored everything. Far from it.

Theodora followed a path taking her over a bridge when bumps formed along her skin. She sensed an external force boring into her neck. Her senses whispered of something amiss, the sensation of being followed. She faked interest in the items on display around the shop poles and in windows, but let her mind focus on the sounds around her. She concentrated on the soft thuds of boots around her; some a slow shuffle, some a forced stutter as they tried to hurry around the bodies. She found the set that was slowly picking up speed and she tugged her dagger out, aiming for where she imagined the person's neck to be. Her judgment was slightly off, the blade a few inches shy, as the man who stood before her was much taller than every other Seclusian walking around them. Her eyes met his chest. He had not even flinched in response to the blade.

Maddox.

She forced her face to remain neutral, not giving him the satisfaction of her errors, but she knew she was off. And she knew *he* knew she had missed her mark. Her lack of visits had dulled her instincts, for training usually failed to prepare one for reality.

"Theo." His voice traveled down the blade.

"Dora," she grounded out., "You know it's—"

"Oh no, have you already forgotten? The name's Maddox." He feigned innocence, but a smirk crept across his lips as he buried his hands deep into his vest pockets.

"No, I meant..." A growl of frustration harmonized with her retort.

"I love when you get flustered."

"I'm not flustered. You just—"

"I just what, charm? Irritate you beyond belief?"

Charm. So, nothing had changed. She wanted to tell him he was right; this was why she hadn't returned to Seclus sooner. But she wasn't sure if he would know the statement would hold more lie than truth.

Of course, her lack of response didn't appear to puzzle him. He took the silence in stride, not letting it break his banter. "To what do the good people of Seclus owe for you to grace us with your presence once more?"

She couldn't help the warmth creeping up her cheeks. She didn't know if he would've even noticed her absence, but she shouldn't have been surprised. He always noticed more than he should. She ducked her head for a moment, attempting to force her face back to coolness as she secured the dagger back to her waist.

"I'm only here to brandish my blade against the average commonfolk."

On the contrary, there was nothing average about Maddox. It wasn't merely his height, but also his darker than black hair and his dark eyes. He appeared both out of place and out of this time. His clothing was pure elegance, and his neckline typically donned some type of cravat of rich red, a fashion statement which had failed to catch on in either of the cities.

"Average commonfolk? You wound me with such harsh words."

"I wound nothing. If you must know, I'm headed for Hakon and his possible job. I would ask if you want it, but I've already been told, since your fellow Seclusians ratted you out, informing me you left it for me."

"No such betrayal exists when I tell people to inform you. I wasn't sure you would get word up there in your beautiful capital."

Theodora wanted to say she knew and wanted to tell him she knew a lot of his ways, more than he probably realized. He was becoming predictable, and she needed space. Space away from him to sort through such weird feelings that were taking root deep inside her.

"I'm headed in that direction," Maddox continued, offering his arm out to her. "If you don't mind, I'll walk with you."

Instead of a softness in his eyes, which she was hopeful for, she was only met with his bold gaze, daring her to tell him no. Arrogant bastard. She reluctantly intertwined her arm into the crook of his as they continued along the bridge.

The questions in her mind were immediate. *Why was he here? Why was he walking with her? Why did he offer his arm?* Yes, among the Seclusians, their names had become inseparable, their reputations knitted into exaggerated gossip. But in reality, the two rarely mingled. They didn't work tasks together or make elaborate raids or grand plans. They each acknowledged the other existed from afar; a wink or a smirk across crowds, all from Maddox of course, or her own nods in acknowledgement, but never anything more.

"What do you want, Maddox?" She was unable to keep the curiosity locked away. The bite in her tone wasn't entirely unwarranted.

"Have you learned of the recent Lawless attacks happening in Conlis?"

Dodging the question, as usual. She rolled her eyes.

"I have. I've also heard they are moving closer to Lume. What of it?"

"It feels different. Like the world is on the brink of change."

"Let it change then."

"You're that indifferent to change, charm? You don't wish to be a part of it? To either help bring the change or to prevent it?" He waved an indignant arm around them. She knew he was reminding her of all the *changes* around them. "It's certainly a surprise. I thought it was why you became a Lawless."

Theodora stared and Maddox only smirked back at her, raising an eyebrow for her to answer the question.

She shook her head. "No. I only do *all* this for klaud. Nothing more. Do I always agree with the king and his Satelles? No. But it's why I'm one of the rare Lawless."

"Rare only to Lume, charm."

A small chuckle escaped her lips, and she couldn't stop the smile this time. And this was why she couldn't stay away. With all the chaos and threats, she was forced to stay thinking a few steps ahead of the next person, but with Maddox, it faded away. She could let her guard down for a moment because she knew he would take care of it for her. Even if she never received his entire story, because she was sure Maddox was filled with hidden secrets, she knew he would still watch out for her. And unfortunately, that was a singular exception for her.

They followed the paths silently. Theodora noted the differences between her capital and this city. A city full of superior tinkerers. All persons here wore tunics and trousers with pouches of tools and holsters with weapons. While the work here appeared to be unending, Seclus had an ease to it. The sounds of the pulley systems, the grinding of gears, and the metallic clangs were an orchestra, and the people were the musicians. Maybe instead of blood filling their veins, there was fuel instead.

As they continued further into the depths of the world, Maddox moved his left arm into his vest pocket, fishing something out. Theodora's body tensed in anticipation. The shine of metal caught her eye, and she noticed a chain ending in a thick circular disc. She tried to lean closer to garner a better look. It appeared to be a pocket watch. Maddox displayed it for her, a silent answer to her curiosity. When the cool rounded metal slid into her palm, she unwrapped her arm from his, stopping in the middle of the path, the chain still connected to the inside of his vest.

People bustled past them. The faint smell of petrichor and an ancient indescribable scent filled her nose and mixed with the recycled air from the ventilation system. A small breeze brushed past her cheeks from one of the ceiling vents. This was a collision of the fates meeting new technology.

Theodora turned the pocket watch over within her hands. It was exquisite. She stared at the bright silver that now gleamed at her. The intricate metal was crafted into a spider web on both sides, the holes in the web showing glimpses of the face of the watch on one side and the gears on the other.

"It's beautiful." It was a whisper that escaped her, as if even saying the words too loud would allow for the watch to crumble within her hands. She ran her fingers along the cool, delicate piece.

"The most beautiful I've seen."

She jerked her face up to him, finding the weight of his gaze heavy on her. A heat rose from her core, moving up her gut, squeezing her chest. She attempted to force it back down as she returned her attention back to the creation, chastising herself for such thoughts. These feelings weren't foreign to her, instead steeping her mind the past petriks, unwarranted and unbidden—actually, unwarranted wasn't correct. It was a longstanding desire, which she wouldn't allow herself to acknowledge. She had hoped a break from Seclus, a break from him, was enough. But even Danika's overbearing personality had been unable to squash it.

She returned the watch to Maddox, before continuing along the path again. Maddox clicked the top of it, forcing the front to pop open, glancing at the face before closing it deftly with one hand and returning it back to his pocket. Watching him was unnerving. His confidence and sheer arrogance... Intimidating was an understatement.

Maddox slowed at the divide in the path, one branch continuing straight for Hakon's dwelling, the other turning abruptly to a set of stairs leading over a bridge. "Well, I needed to be headed this way as I have a prior appointment I must attend to."

"Maddox, I don't need an escort through Seclus."

"I know you don't, charm. But it gave me absolute pleasure to do so." With a wink he turned for the stairs, headed for whatever prior engagement he had. Theodora couldn't stop her mind from careening down the path with him, her thoughts growing darker with each of his steps. Maddox meeting her was odd enough, but to escort her here with no other reason?

It would mean the worst for her. Or someone else.

∴ ∵ ∴

Theodora turned away from the spot and continued along the pathway that led to Hakon's home. It wasn't his actual residence, but more of a business location for meetings. The cottage-like building was faced with dark, ashen gray stone. The window boxes at the front were crowded with night gladiolus of deep purple, the spicy scent reaching her nose and mingling with the soil. Even in conditions that were both dark and fierce, life found a way to endure, although the fiedelight bulbs Hakon's wife used daily certainly helped.

The wide door greeted her, and she knocked, the thuds of her fist vibrating into the wood. Within a few moments, a young boy with unruly blonde hair cracked the door open.

"Ye', ma'am?" The sounds fell through the gaps of his missing teeth. He appeared nervous, which meant he was new to Hakon's employment. The older ones were more familiar with the expectations and knew who she was.

Theodora smiled at the boy as she dropped down to his height and whispered, "I'm here for Hakon. Tell him it's Theodora."

At the sound of her name, the child tensed, and his eyes widened in either amazement or disbelief, it wasn't clear which.

"The-The-Theodora? Like the one—"

"Yes, lad." She rose but continued to wear the smile for the boy. "Now go fetch Hakon, as my patience is wearing thin."

"Oh, yes. Yes, ma'am."

He scurried away, unsure of where to keep his attention, either on the Lawless at the doorway or where he was walking. After tripping over the rug and having his face almost meet the floor, he flustered into the home.

Sighing inwardly, Theodora stepped over the metal threshold, squatting down to fix the intricate woven juniper rug on the floor before shutting the door. Straight ahead was a long hallway which raced to the back of the house. The steps to the floor above peeked through the shadows. To her immediate left was a closed doorway; to her right, further down the hall, was an open archway. The spice of the gladiolus loomed in the hall, now mingling with the added scent of must, dirt, and old papers.

Moments passed before Hakon came into view. He was a rounded man, dressed in the highest of finery. He wore brown trousers with a matching vest and a crisp white shirt. The gold buttons on his vest were vast in size and numerous. They would have been eye-catching if it weren't for his bright, flame-red hair swirling around his head.

Hakon had earned his status in Lume through his farms and gardens, which he controlled throughout both the capital and neighboring lands. His name became known and earned recent requests by the king: banquet feasts, festivals, or whatever new social event the king believed would distract the Lumens. Striving for more notoriety, Hakon purchased this cottage to prove he could grow in even the darkest of places. It had become his most profitable garden; the window boxes Theodora had noted in front of the building.

Before growing into a man of wealth, Hakon was a man of manual labor. And no matter the fiedations which passed, it was something he failed to erase, try as he might. His belly was rounder with his increased number of feasts, but his build remained strong and muscled. His hair and clothing were pristine, but dirt always spotted under his nails. He had a sense of brightness that hadn't faded, even though gray threatened to invade his ginger beard.

"You scared off my help, Theodora." His voice boomed down the hall with his approach.

"If you had older ones working for you, they wouldn't be so intimidated by stories."

Hakon let out a chuckle, and as with everything about him, it was contagious. "What is it I can do for you?"

"I'm here for the task you requested."

"Ah. I should've known. Well, come on in and make yourself comfortable."

He directed her through the open archway, thumping his way behind her to where one of two wing chairs stood decorating the room, waiting for Theodora to approach the other. She swept her eyes over the room, her gaze catching on the stelgladio leaning in one of the corners. *Fates be damned.* Her mind tumbled through the possibilities of why Hakon would have one. Although he was appreciated by the king, these were reserved strictly for the Satelles, and to go undetected by Seclus—she needed answers. Theodora shook her head as if to clear the thoughts and forced an emotionless expression before falling into the pale green chair. She laid her arms over those of the chair, bringing up an ankle to rest on her knee. Hakon wedged himself into own chair, losing his calm demeanor when he was forced to remove a couple of throw pillows out of his way.

He cleared his throat. His own hesitation was palpable in Theodora's mouth. It tasted forced and of something else she couldn't place.

"It's quite a simple task. No—" he gestured his hand, accenting the word, "*intended* murders."

Although Theodora had never obtained a task from him before, this wasn't uncommon. With the death of the Lumen queen, their king faltered. There was too much room for error even with the amount of roaming Satelles. With Seclus, the Lawless had found a place where shadowed meetings and hushed conversations were becoming the norm, even welcomed.

Hakon paused again, confliction jutting forward with his eyebrows. Theodora didn't fail to notice the droplets of sweat starting to bead along his ginger hairline. "The task is to enter Amicus' dwelling undetected and locate a safe. It's my understanding it's tucked away in his closet. Inside the safe are a few documents I need containing information with respect to my mother's death."

Theodora's eyebrows pinched together. She knew those statements were a lie.

"You want me to break into a Satelle's dwelling—which, let's cut to the chase, is located in the castle—to steal back some useless pieces of papers?"

"They are not useless," he grounded out.

"Okay, then what do these documents entail that would require breaking and entering?"

"It's none of your concern," he responded, an unexpected fury meeting his words. "You're strictly a hired hand. You get what I request and get payment. That's it." After a breath, his anger cooled to apprehension. "Please. Can you do it without questions?"

A chuckle escaped Theodora. She glanced across the room to the stelgladio, curious as to what role it had with Hakon and maybe even this task. She met his gaze once more.

"Hakon, Hakon, … Why don't you save us both the trouble and tell me what's really going on? And how about we start with why there's a stelgladio in your home?"

Theodora had learned their creation was ordered by the late King Rufus to counteract the shineguns. A blast from a shinegun, when properly charged, decimated the Satelle's old blades. Stelgladios had a magnetic-like current running through them to deflect a shinegun blast. They were like their ancestors in their handles, weight, and balance. But their metal was black and an electrified thrumming could be detected when held.

As much as Hakon may have thought Theodora would believe this task involved some inheritance, she knew he was acting foolishly, which meant someone else was already behind this.

But her questions were met with silence.

She heaved a sigh, placing both feet on the ground as she slid her elbows down her thighs, interlocking her fingers together in front of her chin. "I didn't get my reputation by blindly going into tasks. Something you feel is unimportant can be the tipping point for my success."

He stared at her. The conflict was clear. She stared back.

He forced the words to follow, dragging them from wherever he was keeping them hidden. "I've been told, by a credible source mind you, that Amicus has some documents detailing who may have been involved in my mother's early death and *could* have some information about another unknown heir to the estate. So, as you can see, this would be my interest in it."

"Thank you. I will allow you to keep the source in confidence, as we all need our secrets. And for payment?"

"I've offered four hundred klaud to start and one hundred more after completion."

"Fine, consider the task done. And what of the stelgladio?"

"It's from my source. And they require you to bring it with you when you head to the castle. They said to leave it in Amicus' room as a message."

"You want me to take a stelgladio and walk out of Seclus with it?"

"Are you Lawless, or are you not?"

Even the fates appeared to hold their breath.

"Give me another five hundred klaud and I'll walk out with it right now."

"The blade is part of the task and the initial five hundred."

"I don't know which fates you bargained with to get it here, Hakon, but I'm not walking out of here with it strapped onto me without some compensation. Consider it hazard pay."

Theodora didn't receive an acknowledgement, only Hakon pushing himself quickly from the chair and out of the room. His boots were an echo of what she had agreed to.

She refused to look at the blade, but instead glanced about the space. The walls were carved from the ground with two window openings. No glass fit the openings though, as there were no threats or storms or seasons here. And the unspoken rules among the Seclusians kept each other at bay. Most of the time.

On the wall behind Hakon's chair, a piano held countless pictures of Hakon, his wife, and their six children over the fiedations. Some were clearly new photos that had been placed into their frames recently, their lack of dust revealing the age of the others. The pictures were filled with laughter and smiles, pulling Theodora into their own happiness. A sense of longing crept over her skin.

She stepped out of the chair to see what lay behind her and stopped at a work desk. The light oak presented several tools, drivers, and screws as well as shineguns, dumguns, and flash crystals in various states of assembly.

Hakon's boots announced his presence, and she approached the archway and the corner. He held a leather pouch out and she took it, her fingers grazing his rough hands for an instant.

"Five hundred now. Five hundred when it's done."

She looped the pouch onto her belt as Hakon moved to pick up the stelgladio, presenting it to Theodora. She didn't allow her emotions to show, forcing them to stay locked up deep within her gut.

She grabbed the scabbard and noticed the double strap system. Whoever left this knew she wouldn't have a harness to carry it. But it was peculiar. She never saw any of the Satelles using a system; they always allowed for them to hang at their hip. The questions continued to bubble up as she put her arms in, the stelgladio resting along her spine.

Hakon cleared the archway and gestured for the door, dismissing her abruptly. Uncertainty made for an awkward moment between them. She wanted to question it, to beg for answers. But Theodora had done this for too long. She knew with due time the answers would present themselves, and she would be prepared when they arrived. He reached around her to open the door.

"A pleasure as always, Hakon."

He slammed the door, wood begging the metallic hinges to bend. But like those hinges, Theodora steeled her emotions. The walk was long, and would feel even longer with a Satelle's blade strapped to her.

∴ ∵ ∴

The stelgladio buzzed with energy along Theodora's spine, and she wished she had a cloak to hide it from view. She could feel every vertebra that rubbed against the scabbard. But there was something else, something she couldn't place, as if an itch had formed deep within a fissure, unable to be scratched.

Being distracted by her thoughts, she barely recognized her name being shouted through the tunnels. She turned, the cavernous walls distorting the voice's initial origin with its echoes as she searched for the source.

"Mekari!" She shouted in excitement, embracing when their distance closed, his facial hair tickling her neck. She stepped out of the embrace, keeping her hands on his shoulders. "Look at your hair! It looks wonderful."

The last time she had seen him, he had small auburn curls but now strutted midnight black, making his blue eyes even brighter against the contrast.

"Thank you!" He exclaimed, combing fingers through the curls. "What are you doing here? I thought you had finally retired when I hadn't seen you in a while."

"No, I was taking a needed break, although I'm not sure how much it helped."

Mekari's playful eyes danced at the sight of the blade. "Are you one of them now?"

"Oh, no." A forced chuckle released. "It's part of a task. And let me tell you, this is a first."

"I'll bet it is."

"You don't happen to have a cloak or rag or something I can use to cover it with, do you?"

He patted himself down, baring his empty hands to her. "Sorry, I was only headed back home to pick up some tools Tomas forgot. Speaking of which, don't be a stranger. Tomas wants to have you over for a meal soon. There's lots to catch up on."

"Oh, sure. Absolutely."

"Alright, I've gotta run," he said, thrusting an elbow in the opposite direction. "But seriously. Soon!"

"Promise."

As she made her way back to the markets, she noticed a commotion where witnesses had started to gather. The lighting from the chandelier elongated the shadows from the buildings, obscuring the people in darkness. The ruckus was in rhythm with the stelgladio, making her unnerved and off balance.

Theodora forced herself through the onlookers, attempting to keep her mind focused on leaving, when she caught a pair of glaring eyes nearby.

Jove stood with his arms crossed, his hair dripping down over his shoulders and chest. It was an ice white at the top, fading to green and blue at the bottom. His already angry eyes turned more violent as he marched to her, stopping short of their bodies grazing each other.

His slurred voice was loud, forcing people's attention in their direction. "What the fuck are you doing here, Theo?" The smell of beer poured off him in waves, forcing her to steady herself against the nausea.

"Hello, Jove. It seems like you've forgotten, but the name is Theodora. It's such a pleasant surprise to see you". She iced the words with so much sweetness even her teeth hurt.

"Shut it. I told you I never wanted to see you again."

"Then why stop me here?"

He stared at her, his chest rising and falling quickly. She was sure his lungs were breathing resentment rather than air. Jove was an abusive sort, for whatever reason she didn't fully know. She wasn't sure if it was from his own horrid past or jealousy, or even because he felt threatened by her own title of Lawless, but he attempted to gain control of his fellow Seclusians with the rule of his fist. His friends of sorts stood around him now, slightly aback for their advantage.

"I'm in charge down here. I told you to leave and then you show up creating more issues."

"In charge?" Theodora's laugh resounded over the growing quiet of the crowd. It was clear whatever the commotion had been occurring before had dissipated, and this was the new interest. Theodora sensed the stares around them and noted the limited space distancing them from whatever this was. "Only one of us is creating issues and it surely isn't me."

She paused for a moment. "Jove, seriously, you're drunk. Go home, sleep it off, and we can discuss this tomorrow." She attempted to turn away from him when his hand wrapped around her upper arm.

"I'm not done here. Why do you have a Satelle blade? Are you working for them now?"

"You're an idiot."

"Should I yell *spy*? Tell Seclus you're a traitor?"

She tugged on her arm, pulling it free, but he only moved to grab her other wrist.

"You're going to have to try a lot harder than that. I think restitution is in order."

The stelgladio was incessant and echoed her annoyance back at her. She took a deep breath through her nose trying to forget it. But it bore into her bones. She wanted to be angry with Jove for being a further annoyance, but she knew the beer was in control right now.

"Jove, come on, let me leave. You can go back to your men, have a few more drinks, and maybe find someone to warm your bed."

"Like I said, a lot harder, Theo," he whispered before kneeing her in the gut, forcing her to the ground. At this point the crowd was too involved, craving a fight for their patience, and with the traitorous blade on her back, she was an easy target.

Jove lowered his head, squatting before her as she coughed out the pain. "Now tell me why you have a Satelle's weapon down here."

Her anger surged. She used her momentum, grabbed the sides of his face, and pulled. Her skull struck his face, and she was met with the satisfying crack of his nose. She rose, pulling her dagger from its sheath, and took a step back.

His hand went to his face, now covered and dripping in blood, painting the stones with deep colored spots. His men were unsure what to do next. As Jove rose, he continued wiping his face, looking at the blood in disbelief. Although adrenaline pumped through her veins, her breathing remained steady. Sure. "One. The name is Theodora." She took a couple steps back, bringing her closer to the exit. "Two. I'm leaving now. When you've sobered up, we can talk about how you made a mockery of yourself."

But as she attempted to back up, she ran into people. The crowd was much thicker now, swarming like bees to honey. She calculated her escape, knowing it would be more difficult trying to get through the bystanders, especially if Jove's men joined the fight. Jove laughed, blood still dripping down his mouth and chin, covering his tunic.

"You think I'm letting you leave after that little stunt?"

She saw his intentions in her mind, as if they were her own. The liquid drug refused to loosen on him, keeping him wholly obsessed on whatever this forced interaction was meant to be. Theodora flipped through her options, knowing she had speed and dexterity to her advantage.

Jove moved, attempting to barrel at her, when she saw an orb a split second before an explosion erupted.

On instinct, Theodora brought her arms up, protecting her head and face. A bright light flared, blinding everyone. A flash crystal. Immediately, cries of surprise became a dull echo in her ears. She blinked back the light and saw others starting to run, dispersing from the situation.

Theodora rubbed her fingers over her ears, trying to stop the ceaseless ringing. She opened her jaw, trying to force her ears to pop in hopes anything would bring relief to the noise. She stumbled, realizing her foot had caught on something. Looking down, she saw Jove lying on the ground, his abettors rooted in place with surprise, small burns mutilating their hands and faces.

She glanced once more at Jove's motionless body. His tunic was spotted with blood and glass shards. She knew what his face would look like, but it still didn't stop her shock at seeing his skin burned, in some places down to the bone, and the glass freckling his face.

Condition

The pain in Maddox's left hand was close to unbearable as he squeezed the second flash crystal. His mind raced as his eyes traced over Theodora's body to confirm she was without injury. Her green eyes found him as she recovered herself, and it was almost enough to disarm him.

His attempt to keep his emotions locked away had been unsuccessful. They had broken free and, in that moment, the first flash crystal had left his hand in response. He had meant to only confirm she had gathered the stelgladio and then intended to slip away unnoticed.

He watched as three others nearby patted the burns now coating their skin, glass embedded deep into their flesh. Disbelief shone in not only their own eyes and Theodora's, but also in those of the Seclusians who hadn't immediately run. Theodora's continued stare in his direction weighed heavily on him, but he refused to acknowledge it. Instead, he turned on his heel, slipping the second flash crystal back into his pocket and heading for the surface.

The commotion of the people followed him. He knew Jove was dead. The flash crystal had found its mark effectively. Maddox didn't know the extent of the injuries to the others, but he didn't care to know either; only to ensure that Theodora was alive.

He heard her boots before she surpassed him and stopped in his path in the rocks' shadows at the entrance to Seclus.

Theodora crossed her arms angrily, glaring up at him. The slight breeze from the topside caught her escaped caramel locks and dragged them across her face.

Maddox bit the inside of his cheek to prevent himself from laughing at the situation. Her frustration was contradicted by her stubborn hair. He leaned against the rocks, and a smirk formed out of habit. Within an instant, her hand slapped across his face. He turned away from the tingle in his left cheek before looking back to take her words next.

"What is wrong with you?! I don't need someone helping me fight my battles."

"I wasn't fighting your battles, Theodora."

She moved again quickly with her right hand, but he caught her about her wrist. "I allowed it the first time. I will not allow it again." His voice was dangerously low, the unspoken threat clear.

She ripped her arm away, and he only watched as she mounted the horse nearby. He was not inclined to stop her and had no internal debate about trying to make her feel at ease. She raced off into the distance.

Her and her horse's silhouettes grew smaller against the pink and yellow entanglement in the sky above. He turned his face south, the half slice of Petram awakening at dusk. He studied it to find those hints of pale greens that could be spotted on extremely clear nights. Tonight wouldn't be one of those, the clouds already racing to blot out the setting sky and Petram.

Maddox returned to Seclus, the chaos from his earlier actions already settling. The crowds had dispersed, and some of the local tavern owners had staff outside with samples of food, trying to get the people focused on other things.

The Seclusians were getting restless. They had traveled here with the expectation of making changes to Lume, not only with their technology but also with their leadership. However, he knew Seclus had gained its position not through rushed impulses, but rather through patient decisions. That was something he had forgotten when he threw the flash crystal. He hadn't wanted to create a strain between him and Theodora. It existed well enough on its own. He was certain her stay away from this city wasn't for lack of tasks or even because of the Satelle she had pretended to court. He knew it was about him.

"Sir." Mekari removed his hat in a brief bow before continuing. "Valix sent me to inform you they got the others apprehended. He assumed you would want to speak with them."

Pushing aside thoughts of Theodora, he followed Mekari to the detaining room, focusing on what he could control. The room was tucked inside a cavern behind Previt. It had been one of the first caverns where Valix and he had played as children. It had been part of one of their first adventures here, one of their first tasks together.

The three who had been clearly supporting Jove were at the mercy of how active the glow creatures were, their light ebbing occasionally. Right now, they were mostly lit, their pale glow reflecting off the group's white clothes. The three men were held within separate charged cubes, the electric currents like that of the stelgladio blades.

Maddox walked before each cube, staring down the individuals. He didn't need to say anything. All Seclus knew who he was. "I'm going to ignore your acts of defiance against the Committee," he said slowly, deliberately. Letting each word carry a silent threat. "On one condition."

Opportunity

Entering the outskirts of the capital, Theodora dismounted the mare, who followed along lazily behind her. The sky was a deep violet as Fiedel dropped below the tree lines. The night was a blanket draped over her, giving her opportunities to hide away from the few Lumens as they raced home for dinner, or from the mindless wandering of the Satelles.

A bulky man eyed her eagerly, no doubt curious about the stelgladio. Theodora tugged Down River into one of the alley streets before pressing her back against the wall.

Crates and boxes lined the buildings, the occasional trash or unwanted items littering the cobblestones. She spotted a cloak tucked between two weathered boxes, holes dotting the fabric from moths' prior delight. Tugging it free, she threw it over her shoulders, ensuring it covered the blade. The musty smell of forgotten fabric mingled with a hard scent of smoke. Its previous owner's life had been filled not with cigarette smoke, which she occasionally noticed in the taverns, but something richer she could only assume was the occasional ginseng.

She wanted to focus on anything else other than the rattling inside of her head, but it was an endless back and forth, like a duel in her mind, between Hakon's task and Maddox's intervention.

With her restraint waning, she tugged on one of her pouches in front of her, reaching inside to locate the drystone. In the failing light, she forced herself to watch the scrape along her dagger's blade, letting each pass of stone and metal push away the memories from the surface.

A lot harder, Theo.

She jerked her head, but it was like an annoying bug that refused to leave. It shouldn't have bothered her the way it did. Death was no longer a surprise. She'd witnessed it firsthand with her own parents. But maybe it wasn't death that bothered her, but rather Maddox.

Theodora didn't understand why he had interfered. He rarely made decisions on emotions, using only logic. And she was unable to determine where his logic lay in becoming involved. Although they respected each other, it was out of their mutual value to one another because of their profession. Their work, their tactics, their endgame remained theirs alone. Which only left the implication that emotions had been involved—proof positive something was amiss.

The push and pull of her sharpening blade helped to numb her mind, diverting her focus away from the politics and menial games she didn't want to participate in.

Packed vendor carts wheeled past on loud, rickety wheels pulled her from her thoughts once more. The savory scents of food wrapped around her, reminding her of an old stew her Grandmama would make when the astrum chill would threaten to take over solta, and of how little she had eaten today.

She shoved away the blade and drystone in frustration. A chilly breeze tripped down the alley as she inched to the corner. The line of vendor carts wheeled to their residences, leaving but a handful of Lumens behind in the square.

She skipped her eyes over the people as Down River nibbled at the fabric of her hood. Theodora finally spotted them: the handful of Satelles patrolling the area, their hands pridefully on the hilts of their blades. She pushed the mare back, further into the alley, deeper into the darkened shadows, before creeping closer once more.

"Do you want to go to Rosa tonight?" the shorter one said.

"Why would we go there? We always go to Digere. Albani is much more open to Satelles, even giving free drinks when people aren't being too rowdy."

"I've heard Rosa has some good wine."

"Where are you from? Vindem? We have nothing here except the king's beer, at least until the festival."

"Seriously?"

"The joys of living in the capital instead of the notorious vineyard."

"Will Amicus be there? I wanted to meet the Satelle Captain. The stories of his youth inspired me to come to the castle for my oaths." The shorter one raised his head further, attempting more regality in front of his comrade.

"Yes, he should be there. But, wait, what stories?"

"The one with the prince's revolt against the king? With his hand?"

"Oh, you mean…"

But Theodora had stopped listening, the words of the Satelles fading. She thanked the fates for the opportunity. Even though she wanted a warm meal, this was her chance to head to the castle.

Enough

Danica and the Satelles had provided Rajveer with limited information when it came to this Lawless, Theodora. He didn't know what to think or what to *do* with the few secrets they had shared.

He knew he had to be patient. Maddox was off placing clues to get Theodora to come to the castle tonight, and he was stuck here to ensure no one caught onto what they were doing. And worst yet was this intrigue which had begun with the stories Danika did share. He had always been curious about the young girl who had escaped the Satelles. Now she had a name and a reputation. Meanwhile, in the same amount of time, all he had managed to do was lose his betrothed, his mother, his hand…and be forced to deal with another night of this shit.

Dust flew up from his book as he slammed it shut, tossing it down on the wooden table next to him. The book thudded against the oak, echoing his annoyance. Rajveer wasn't much of a reader, but his mother had turned to books during difficult days, pulling out the stories after Rajveer had trailed his father all day. He had hoped a book would help calm him tonight. But even the lavender scent that clung to the book, a story his mother had read him countless times before, was unable to soothe his nerves.

He should be doing something. Anything that might help the terrible task which now lay before him. He still hadn't discussed his father's decision to stand down during the Lawless attack this morning. It was exactly what his mind needed to focus on right now.

Rajveer tugged a cigarette box from inside his jacket, withdrawing one before striking the lighter. He lit it and returned the box as he pulled in a deep breath. One of the castle staff hovered near the door exiting the library, and Rajveer caught her attention, waving her over. The light from the lamps shone around the dust particles which swirled about as she hastened to him. Her citrus perfume reached him first, and he blew the smoke past his lips to mingle with it.

"Where is my father?"

"He should be finishing up his meal in his study, sir."

Rajveer remained silent, letting the momentum of the deep pulls of the herb slow him down for a moment and contemplate whether he wanted to have another fight with his father. His only other option was to sit and ignore it, or find something else far more worth his time.

"Sir?" The woman hesitated, the silent reprimand hanging first between them before she vocalized it. "You know the king doesn't like smoking within the castle walls." She gave him a small smile. "Drop the butt into the water pitcher and I'll have it disposed of." She turned to walk back to her post, leaving him once again to his thoughts.

This was the dilemma. He was stuck in a world where his staff adored him, his Satelles followed him, and the citizens praised him. But he was shackled to the crown, to the king, just as everyone else.

He took the last few hard pulls, savoring the bitter smoke as it filled his mouth. He crushed the end with his metal fingers and dropped it into the pitcher. He rose, tugging the jacket closed around him and looping the button into place before nodding to the woman.

Fates, grant patience.

He left the library and entered the cold hallway before hiding his left hand in his pocket and forcing his feet past his parents' previous suite. Even after all this time, his reaction remained instinctual. He acknowledged the Satelles he passed as he moved to the spiraling staircase. Although the castle had been made of harsh stone, his mother had tried to make it welcoming with ivory drapes billowing in the archways.

Rajveer made it to the western wing, where he interrupted his own thoughts with a series of knocks on the large door. Before pushing it open, he reminded himself why he was here. Maybe he could persuade his father to relinquish some of this control. With his father surrendering some power, maybe Rajveer could reevaluate this Lawless plan.

He found his father hovering over his nearly empty plate, a chicken bone held precariously in his weathered fingers. Rajveer took a moment to realize exactly how much his father had aged over the past few fiedations. The warm brown of his hair was peppered gray from time. And although Rajveer had inherited his blue eyes, his father's were now filled with a sadness that stole their beauty. If he were told stories about mythical blood-drinkers that came in every night to feast on his father, Rajveer might have believed them.

His father's study was scattered with papers, tucked into free space and crevices. They were taxes and profits, and correspondence from neighboring cities. The bed was made and meals cleaned up daily by the staff, but his father wouldn't allow their services to extend beyond that.

His father continued his meal, licking and nibbling the bone clean of all its meat. He hadn't even acknowledged Rajveer when he knocked or entered the room.

"We need to discuss these attacks, Father. We need to get the Lumens back in our good graces. No longer living in fear, but trusting in us."

The king merely grunted in response, licking the last of the chicken juices from his lips before gulping his wine and slamming down the glass.

"Father, please," Rajveer tried in a calm tone as he slowly approached the table. "We can unite our people, bring them together and remove Seclus completely—"

"Are you king?" his father bellowed, his voice reverberating off the stone walls. He threw the bone onto his plate and tugged the napkin across his hands. He sank back into his chair. "Have you found a wife?"

"You're being ridiculous! You won't even consider this might be best for our people..."

"Our people? I told you how to earn this crown, Rajveer. I told you when your mother died that this mantle was waiting for you. All you needed was to find a wife."

"That's it? That's all—"

"That's all. And until then, you are going to keep your mouth shut."

"Seclus is going to take over the capital!"

"We have no reason to believe these attacks are coming from Seclus. The Satelles themselves couldn't even touch me if they wanted to. And furthermore, if Lume does fall, it will be because of you. You will shoulder that burden."

Rage blinded Rajveer—how could his father be this irrational? And over something so ludicrous.

"Why do I need to find a wife? Why is marriage the contingency? The only contingency?"

"I won't speak of this again, Rajveer. You find a woman to marry, or you will bear witness to this capital's destruction."

"But this is—" Rajveer tried again, moving closer to his father, hoping he would see his desperation.

"Enough!" His father jumped from his chair, causing it to fall back with a loud thud.

"If we tried—"

The tightness of his father's hand on his throat cut off his words. The king pushed him back across the length of the room, and Rajveer struggled to keep himself upright, until his back was pressed against the door. Rajveer felt each fingertip as it pressed further into his throat, removing his ability to breathe. His father's breath was hot and sweet against his cheek.

Rajveer didn't want to retaliate; he knew it would only make it worse for himself. He had already begged enough for his people. He wouldn't give his father the satisfaction to do so now for his own life.

But with his last bit of air, Rajveer was forced to whisper out, "Father, please."

When his father released him, Rajveer immediately fell to his knees, touching his throat with his metal fingers, their chill washing over the already sore muscles in his neck. He gaped up at his father, but the king had turned back to claim his seat at the table, the chair already replaced to its upright position by nearby staff.

"You'll regret this." Rajveer's voice was hoarse as the words left his mouth.

"Will I?"

Rajveer stood, containing the retort. As much as he wanted to have the last word in this moment, he knew, in the end, he would.

Undetected

A handful of leaves danced across the cobblestones as a soothing breeze whipped past Theodora. The heat of the day had been vicious, but the slow arrival of night was a gentle reminder of the approaching astrum. She tossed back the last of the beer from the skin she had purchased from one of the vendors. Although his cart had been packed, he remained eager to serve her food, the possibility of a few klaud able to fight back some fear of the patrolling Satelles.

She walked the outskirts of the capital, headed for Silva Forest, which divided the capital city from the castle. She still didn't understand why the castle was so far removed, but shoved those thoughts aside, digging into one of her pouches to find her fingerless gloves. They did little against the chilled air, but the sides were laced with metal, another weapon to be used if necessary.

She mounted Down River and pushed her south into the thickening woods, choosing to avoid the main path used by the king and the Satelles. As they moved through the felled trees and twisted branches, her mind wandered back to Maddox. She tried to focus on what his endgame was, but instead it fell back into their history.

Theodora had only been a child, a few fiedations after she had run away from the throne room and the king. After petriks of being alone, stealing food when she could, a drought assaulted their lands. As she walked the streets, the brutal heat burned against her malnourished body. Without family or a home, and barely any klaud, she struggled to find purpose. She witnessed those who were fed turning their backs on those struggling. Theodora had turned to the fates, beseeching them to take her.

At some point she'd only wanted to rest against one of the walls to waste away. She was curious if she curled up there, if anyone would even stop to question the dirt-covered body of a child, thinned by lack of food.

Having found no reason not to, she threw herself against the stone, begging for rain to arrive so it might cover the tears which had started to fall quickly. At least with the rain, she wouldn't feel so alone. She would know the fates knew of her pain. But the rain didn't come. Instead, Maddox had.

The Satelles were making their rounds, high up on their steeds, giving them the continued power rush they craved. The scene wasn't uncommon, except this time, a young boy, who she would later learn was Maddox, stood yelling at the Satelles.

She couldn't remember the threats or insults the Satelles threw in her direction. She faintly remembered how they had pointed and laughed at her until Maddox pulled a dagger out and started swiping at their horses. One of the Satelles finally dismounted, and although this man towered over the child, Maddox matched his anger.

The man lashed his hand across Maddox's face. Maddox had attempted to block the strike, but the Satelle's force was too much to stop. Maddox fell to the ground, the right side of his face bleeding and marked with streaks of red.

The Satelles turned away from him, continuing their parade as if nothing had happened. The people who had gathered to witness followed suit, giving the young boy nothing but their backs. As Maddox pulled himself from the ground, Theodora's eyes locked on his. He gave her a small nod and she returned it. In that brief amount of time, Theodora connected with him. She had believed herself to be completely and utterly alone, and yet there were others like her out there, making the loneliness a little more bearable.

Theodora pulled her mind out of the past, chastising herself for getting lost in thought, especially with how quickly they approached the castle. Down River trotted beside the gray stone wall surrounding the castle, careful to remain hidden within the tree line.

Theodora jumped down, abandoning Down River in the trees, and slowly entered the clearing to address the wall, which was thrice her height. The building itself was set further back from the exterior wall, separated by stretches of grass on all sides. A few small trees were planted around the terrace located near the western wall, but otherwise the space was mostly open. The near side of the castle only had a small scattering of windows, the massive ones left for the throne room and front entrance.

During her courtship with Danika, she was able to visit here once before. She'd taken the opportunity to learn of the locations of the Satelles' rooms and how they filled the second floor, each wing containing at least twenty rooms each. Although Danika hadn't listed the residents of each room they had toured, Theodora was adamant about learning which belonged to Amicus while ensuring her intentions remained hidden. Any information could be beneficial for a Lawless.

From the corner of the side wall, she counted her paces south, matching what she remembered from her prior visit where the gardens and terrace would be. Theodora pulled out her multi-tool, wound cable, and opening hook from her pouch, securing them together. She locked her sights on a broken stone near the top of the wall and waited for the fates to give her the cover she needed. When another breeze came through, she fired.

The hook shot outward, punching through the weakened stone, and locked into place. A hard tug on the line confirmed it would hold her weight to scale the stonework. She moved to the base, latching the multi-tool to her waist, and started the climb.

Her boots met the stone as she leaned back, adjusting her weight to compensate for the stelgladio on her back and gripping tightly to the cable. The buildings in the capital twinkled with their stones made with crushed gems, but here, surrounding the king's throne, was nothing except drab stone.

Nearing the top of the wall, she gripped the broken stone and slowly drew her head over the crest. She noticed one guard leaning against a weeping tree. He didn't appear to be asleep, but he certainly wasn't thrilled about the lack of activity during his shift.

She slid her left arm over the wall to hold herself while she pulled a tab on the hook, collapsing it in on itself. Giving the cable a hard tug, the line and hook slipped back out through the opening they had created. She made a mental note to remove the collapsed hook and secure it back into her pouch when she wasn't dangling on the side of a wall.

With another gust of wind, she rolled over the wall and twisted her body. Landing on her feet in a crouch, the end of the scabbard striking the ground, she moved quickly to the closest shrub. She scrutinized the nearby Satelle for a few moments, but his attention remained on something in his hands. She crept closer to the guard, the gold of his jacket reflecting the garden lights. She noticed his hair grew long over his ears as he brought his hand to his mouth and began assaulting his fingernails.

These Satelles needed more action during their patrols.

Weaving through the plants, she found that the shades of gold on the small buds of the first astrum flowers barely glowed in the first layer of darkness. She checked the placement of her gloves, closing the distance between herself and the Satelle. She looked around before rising on her toes. She intertwined her fingers and then swung both her hands at his right temple. The metal plating, decked along the outside of her hands, struck true. She pushed the guard's head to the side, slamming it into a tree trunk. He crumpled to the ground as he reached for the pain in his head. She struck his skull again before she bent low to ensure he was unconscious. Another quick sweep of the grounds confirmed she remained undetected.

No one told her becoming a Lawless would be a life of quick glances and tight movements. With another look around, she was climbing the tree. It wasn't a tall tree, pruned to not reach the height of the outer wall, but some of the inner branches still allowed strength for her to climb onto the roof of the covered terrace connected to the castle itself.

If her memories with Danika were correct, Theodora was swinging over the rails of the balcony connected to the training facility, and on the floor directly above was the prince's chambers. For a moment, she wanted to sneak her way up to the top floor, for an official face-to-face meeting she had always been denied. Sure, she saw him from far off distances, but the mind had ways to play tricks within such a space. She hadn't been lucky enough to see him when she was here as a child, although she was certain it was the queen's doing to ensure her son's safety.

The balcony door entering the training room gave a satisfying click as it opened. Pushing past the sheer curtains, Theodora plunged into a vast room, empty except for the various weapons placed along the walls. The lampposts inside were dim crepuscular light, and she forced her eyes to focus on the shadows, ensuring none were anything more than the furniture and weapons she imagined.

She strained her ears to listen for the sounds within the rest of the castle, but the only one was the gusting of wind outside as she brought the door to a soft close behind her. Politics and memories of the castle aside, the interior was as beautiful as she remembered, albeit darker at night and quieter. Almost too quiet. She ensured her hood was pulled and the stelgladio covered as she entered the interior hallways. However, an uncertainty perched over her. She attempted to wave it away, unable to concern herself with small details, even if they were screaming at her subconscious. She would deal with whatever was thrown her way.

Except for that one thing...

A punch to the face stopped her progress. Theodora's hand immediately went to her nose as she glanced up at her opponent. The young Satelle must have been stationed at the corner of the hallway, because she'd never heard his approach.

Even though the pain wanted her to take a moment to pressure it away, to check if anything was broken, she didn't have time on her side. She needed to subdue this guard before he alerted others. He punched again. Theodora expected it this time and stepped back. The Satelle stumbled, giving her the opportunity to place her hands on his shoulders. With her foot on his thigh, she pushed herself upward, swinging her body around to his back. She slipped her arm around his throat, forcing pressure onto his windpipe.

He staggered back, slamming her body against the wall, the stelgladio clanging loudly against the stone. His movements became lethargic, slowed with the loss of air. With his final breaths, he attempted to reach her face, but his youth told her he didn't have the experience, grasping for tactics of survival.

His body became heavy as he lost consciousness, and Theodora slipped out from under him while he slid to the ground. A thud from upstairs told her she was out of time. She dragged the unconscious guard's body to the curtains draped along the window. Although it was a light fabric, she hoped it would provide enough cover. Rushing through Danika's babble in her head, she headed down the adjacent hall for Amicus' room.

Theodora prayed the Satelles didn't have some weird change of rooms ritual routine. She put a toe to the wood and pressed her ear to the last door on the left, but heard nothing but her pulse, loud and thundering. She pushed the door in slowly, the wood barely grazing the stone floor underneath, teasing it to make noise. After slipping in and brushing the door closed behind her, she settled her hand on her dagger as her eyes adjusted to the darkness. Hearing nothing more, she located the chain for the lamps along the wall.

She pulled, the chain clicking as it triggered the lights to jump to life. Darkness pushed away quickly into shadowy corners. She forced her eyes to take in her surroundings. She was in a small study of sorts, full of bookcases, rugs, stacks of books, and a closed door leading to another room. Vision still blurred from the shock of light, she blinked, trying to focus. She turned to look at the other side of the space, and found a man in a single dark-colored wing chair, his hands steepled in front of his face, his blood-red cravat demanding attention.

Incomplete

Maddox watched eagerly as Theodora entered the room. Sitting in the dark tricked his interpretation of time. Even with the random sounds from the next room, he lost count of how long he'd waited. It didn't largely affect him. Anything pressing would be handled by Valix during his absence, but it annoyed him to be forced still with incomplete matters he needed to address.

"What are you doing here, Maddox?"

The way she said his name. It was like an insult.

"Same as you. Hakon sent me here as well, although he didn't give *me* a stelgladio."

He tilted his head, gesturing for the door in the middle of the small vestibule. It creaked loudly as it pulled open. The room beyond was dark, shadows moving within. Maddox looked back at Theodora. He witnessed the moment she registered the movement, how her eyes widened in disbelief. In the next breath, she stomped her right foot on the ground. The spring-loaded sheath he knew was within her boot retracted, her dumgun shot upwards, and her hand caught the grip as she took aim. The filigreed barrel gleamed in the light.

"Let's not be ruled by emotions here, Theodora," Maddox attempted to soothe her. He didn't know if she was still angry about Seclus. Even though he strategized various scenarios before, it was still a rush when the pieces lined up perfectly on the board. As much as she had tried to suppress her expression, he didn't miss the flash of fear in her eyes, like those of a bunny caught in a trap, waiting for the fox to approach.

Three Satelles crossed the threshold into the vestibule, ones he recognized from earlier this morning: Danika, Miles, and Jude.

"Not another step," she breathed out as she shifted her weight to her right foot.

Maddox rose from the chair to move in front of the barrel, blocking her line of sight. He waited as she carefully lifted her eyes to meet his. They were a deep green, like the color of leaves, untainted by people, with no promise of death. They reminded him of home.

"Don't even think about releasing that shinegun," he whispered.

She rolled her eyes—when was she not rolling her eyes at him? She went to speak, but he cut her off as he turned to face the Satelles.

"Why do I always find Satelle meetings are held in the smallest of spaces? Certainly, Satelles could find a better way to talk with two of Seclus' most acclaimed Lawless."

Theodora lowered the dumgun, and he noticed from his peripheral she'd relaxed her stance. It was slight, but enough to pacify him for the moment.

Danika advanced further into the room, separating herself from the other Satelles. "You get what you deserve." But her words lacked ferocity. Her attempt at intimidation weighed down the features in her face, a heat of red spreading over her cheeks. She filled her words with as much confidence as she could muster. "We have matters to discuss."

"No shit," Maddox retorted as Theodora chuckled. "But seeing as Theodora decided to take her time showing up, I have already spent well enough of my waiting. Let's move on. I have other matters to attend to."

Danika's face swept with red. "We gathered you here under a ruse for everyone's safety. Hakon believes we requested a task that would require both of you. However, given this morning's earlier events, I would assume you two are incapable of working together."

"I wouldn't find that to be a safe assumption."

"In any event, Amicus has sent us here to seek out your help."

Theodora laughed. "Can we move this along? I think this is the second or third time that we have asked. What do the Satelles need help from the Lawless for?"

Danika let out a sigh, hesitation stretching among them. It bounced off the walls and cramped the small space even further.

"We've been sent on behalf of Amicus and Prince Rajveer to have the king assassinated. We know Lumens will follow the crown, and we need it placed with the prince. We can shift the power, and hopefully, get both Lume and Seclus on a better path moving forward. If one of the king's Satelles is involved in the attempt, we fear there will be too much civil unrest and power will be lost, creating more chaos and unneeded bloodshed. We'll try to help you as much as we can, but with both of you working together, the Lumens will see it more as a rebellion, and we can help to unite them that way. Bring them hope with peace."

"Impressive," was all Theodora said, as she appeared to process the information.

"A rebellion already exists though." Maddox feigned contemplation as if he didn't already know the plan of action. It hadn't been difficult to persuade the Satelles to act as if this was the first time he'd learned of this. He had made it clear that it would only help solidify her chances of working with them.

"What would Theodora and I receive as a reward for such a task?" Maddox continued. "Clearly this isn't a task for anyone. We would need some recompense for destroying our livelihoods."

The weight of this decision would be pressing down on Theodora's shoulders. This involved her future with Lume, which would be desecrated. Maddox knew that although she was Lawless, her allegiance remained wholly to the Lumen citizens, insufferable king or not. With this agreement, she would be abolishing her home in the capital and the reciprocal relationships she had established here. He didn't know what she would do afterward...

"By doing this," Danika continued, "Amicus has agreed to pardon you of your crimes. You wouldn't necessarily be free to commit more unpunishable crimes in the future, but we would allow certain actions as dictated by the prince alone. Of course, you'd need to change your names and appearances, because there will be some Lumens who will wish you dead." Danika licked her lips before continuing, "Or, with the klaud you would receive, you could leave the capital and spend your days free of all the nonsense. We would help to secure safe passage to Conlis or Nemaaer."

Maddox didn't miss the intense stare Danika pointed in Theodora's direction, but Theodora only remained internally focused, as if she were processing none of the words.

After a pause, Theodora finally spoke. "What of the documents Hakon hired me to obtain? Why do I have this damn stelgladio on my back? Given his reaction during my meeting earlier, he doesn't believe this task to be a ruse."

Danika nodded to Jude, who pulled forth a pristine envelope, sealed with the king's brand, and handed it to Theodora.

"Easily taken care of."

"But what of the blade?"

"No need to worry yourself with it, Theo. It is a mark for the Satelles, and that alone. You can leave it here." As Theodora removed the blade and its system from under her cloak, Danika continued. "Obviously, you will need time to discuss this with each other."

"No, we don't." Theodora didn't hesitate as she rested the blade against the corner wall. "We'll do it."

"Are you sure?"

"I don't need to be questioned, Danika. Besides, this is exactly what *you* want me to do. And if Maddox didn't agree, the conversation wouldn't have gotten this far."

Maddox didn't know what Theodora was hinting at. Her rash decision to agree to this plan was strange to him. There had been no discussion on how they would accomplish all of this. The lack of planning by the Satelles both earlier and now gave him an odd feeling, like mistakes would be made. He had the strange sensation that he must be missing something else if Theodora was eager to agree to this quickly.

"And you, Maddox?" Danika looked at him. "Are you in agreement?"

"Yes."

"Well then, that moves us along quickly. I hadn't even explained our plan. But in any event, Amicus requested we give the king the impression the prince may have found his future bride. We all know about the astrum party, which is typically hosted at the castle prior to the Astrum festival. This fiedation, the prince is hosting the party at the capital house. Unfortunately, the king has no desire to participate in such events anymore." Danika directed her attention to Theodora. "Make yourself present. Let the Lumens see you there with the prince. Hopefully things proceed well enough before the Astrum Festival for the gossip to begin."

On silent instruction, Jude and Miles strode quickly past them for the exit. Maddox would have to tidy up the pieces of the scarcest of plans with Theodora later. Clearly, the Satelles lacked the skills to properly retain their services without raising an unsolicited number of questions. Although the other two had left, Danika remained rooted in place, continuing to stare at Theodora.

Maddox debated staying, to witness the tongue-lashing Danika would no doubt receive from Theodora, but he had other items he needed to ensure were moving appropriately. He forced his eyes not to find Theodora's again before he walked out of the room.

Broken

For the first time in many fiedations, Theodora felt powerless, in a world tilted on its axis and full of a wealth of information she was struggling to process. With this information came the possibility of freedom from expected responsibilities, ones she knew came with being an orphan and ones no person should ever be faced with. She wanted to find escape from the never-ending battle between Lume and Seclus. Now the opportunity had presented itself and her mind reeled.

Between the earlier incident with Maddox and now this task with him, she wanted to tighten the tourniquet and stop the bleeding of emotions. Her past, present, and future were colliding in a short period of time, leaving her to pick up the pieces and attempt to sew them back together.

And now Maddox had left her. His disappearance left her with a chill in the stone room. The chill and Danika.

Danika slowly moved closer, arms out in what appeared to be an embrace. Theodora immediately threw her arm back up, the muscles in her hand still tense from holding the dumgun throughout the entire exchange. Her fingers were borderline numb from squeezing the oak grip.

"Theo," Danika whispered, her voice barely reaching Theodora's ears. "We have a chance now. We just need to get through this. And together, we can—"

"I told you that we were over."

"But don't you see? We can be together and forget all of this, spend the rest of our lives together happily."

"I can't be with you and forget everything else. My past made me who I am."

"Don't be like this, Theo."

"Do you think I enjoy this? I would love to not have to be this way. But there is nothing I can do to change my past. I'll just kill the king, have my crimes be pardoned, and then what? Playing your wife isn't going to make it all go away."

"Please, Theo, let me help fix this."

"That's where you're wrong, Danika. The part you always get wrong. My name is Theodora, and there is nothing to fix, because I am not broken."

Relief

Rajveer was propped against his array of pillows in bed. It was too early to attempt sleep, but with no book able to capture his attention and without the full Petram to bring him company, he remained restless.

His eyes followed the imagery on the ceiling, the muted light from one of the bedside lamps creating shadows from the roughness of the stone. With him squinting and widening his eyes in various combinations, he played out scenes in his head.

Typically, ones with her, his Alouette.

∴ ∵ ∴

"What did my father have to say? Anything of importance?" Rajveer closed the book and leaned back against the pillows to look at Alouette as she entered the room.

"Not really. He only talked about future expectations. You know how he can get," she said quietly as she shut the door behind her.

"Are you excited?" Rajveer asked, the mattress sinking as she sat on the edge of the bed.

"Sure." It was a soft whisper as she picked at the ends of her nightdress, a pale violet further illuminating the coloring of her skin. Her hands had been dyed with elaborate designs, a tradition from her native city of Nemaaer. Rajveer followed the pattern of lines, leaves, and flowers as it flowed up and swirled around her wrist. He immediately looked to his own wrist, the metal of his new hand still feeling foreign even after almost an entire fiedation.

"Are you excited?" She brought her hand to his, covering the harsh metal.

"Absolutely! Tomorrow you will be Princess Klauduisz."

A giggle escaped her, and he brought his gaze to her mouth, where that sound had been created. He wanted to bottle it and drink it in the middle of the night when the nightmares came and he didn't want to bother waking her up.

"You know that's not how it works, Raj."

"Doesn't mean that's not how I'll see you." He rose *from the bed, moving to the table to pour glasses of wine, the sweet aroma filling the room faster than the deep-violet liquid filled the glasses.*

He brought one to her. "In a few fiedations, my father will be ready to abdicate, and he and my mother will live in the capital house, enjoying the rest of their lives together."

"Hm," she mused as she brought the glass to her lips, *sipping the drink slowly.*

Rajveer, giddy for the future, continued once more, "Then you will bear the name Queen Alouette Klauduisz."

"For you, I would bear anything."

"Oh." Rajveer put his glass down and leaned over to *her neck, kissing the softness of her skin.*

"Raj, stop." The words were peppered with small *laughs. "You'll make me spill the wine."*

"I don't care about the wine," he mumbled into her *collarbone.*

∴ ❖ ∴

A rapping at his door tugged him from the vision, and he immediately slid from under the sheets, the soft fabric beckoning him to stay. He slipped into his trousers and left his bedroom and approached his chamber door. The cold hallway air crept into the room as the door gaped open to reveal Amicus' dark green eyes.

"It's taken care of." Amicus whispered quickly. "We have secured them both. She'll meet you at the capital house for the astrum party."

Tension fell from his body and relief flooded over him, if only for a second. The hard part for him was done. Now he needed to continue with the part he'd always played: meet the elusive woman, woo her, and bring her to meet the King. All things he did, every petrik.

But his future was in her hands. The possibility of Theodora's failure snuck into his mind. Although Danika had been hesitant with her information, she had made it clear that by offering Theodora freedom, the woman would rise to the task and ensure Lume had a new king.

"Sir?"

"Sorry, Amicus, I got lost in all of it. There were no issues?"

"None. They offered her a reward that would allow her to leave Lume, and she agreed instantly."

"Okay…great." His brain, still lost in the past, struggled to fully comprehend this new information. "Alright…thank you. Be sure to get some sleep; we can talk more in the morning."

"Yes, sir." Amicus saluted and began to turn away from the door, down the hall to his own room.

"Amicus?" Rajveer stepped into the hallway, propping the door open with his metallic hand.

Amicus slowed and turned. "Yes, sir?"

"Was she attractive?"

"Was who attractive?" Danika turned the corner, an eyebrow raised in their direction.

Amicus looked in her direction before turning back to Rajveer, his teeth peeking from behind a widened smile barely contained on his face. "You'll have to wait and see."

Secrets

Theodora pulled herself onto one of the barstools of the quiet Rosa, a spot along the bar offering both a view of the rest of the bar as well as of the tavern. The occasional glass sliding on wood or scrape of metal along porcelain accented the muffled tones of the couples and small groups dotting the tables.

She caught the barmaid's attention and signaled for a beer. The barmaid acknowledged her, but there was a lack of rush here, a sense that there was no point in racing time as everything would meet its destination, eventually.

Rosa was more refined than most, especially compared with Digere. Mahogany tables and chairs were spread out along the warm cherry wood floors. The floor paneling continued up the face of the bar and the walls of the room. Although the lighting was muted, the perfect atmosphere for the caress of a lover's kiss or the touch of a hand, the space was still welcoming and bright.

The barmaid placed Theodora's beer on a napkin in front of her, and Theodora gulped it down quickly. She never understood why Satelles would drink beers, because she knew the king wouldn't drink such a disgrace of spirits. She felt the faint warmth bloom at her core from the small amount of alcohol.

Only a handful of tables were filled with people as most of the dinner rush had already ended. She shifted her focus away from the earlier disruption at the castle to the current conversations around her. Most were personal, which she quickly skipped over, some were shop business involving deliveries of new orders or supplies, and—hidden within the mutterings—some of the words were far more dangerous.

Just take away his crown and he will fall. Why should he get all the power and the wealth? It would be easy to kill him, if we could get close enough. When's the last time he sacrificed anything? He thinks he can host festivals and we will forget everything else.

Amicus must have heard these rumors and chosen to strike before everything fell to chaos. Lume was on a precipice, teetering between tradition and a new way of life.

King Klauduisz was as secretive as they came. The nation's single mass of land forced cities to work together for resources, and a capital to rule all was formed to provide consistency in their laws, punishments, taxes, and an equal footing to those same resources. When Seclus sprouted from deep within the ground, Lumens turned to their king to keep them safe from these unknown people. But as fiedations went by, he'd only continued as he'd always had: only punishing the Lawless and collecting taxes.

A divide had formed. There was no longer an equal opportunity for resources, because Seclusians had an advantage in technology, threatening not only Lumens' lives but also their livelihoods. King Klauduisz had been a silent king, indeed, as he buried himself deep in the castle, and only digging deeper still following the queen's death.

The barmaid replaced Theodora's beer and slid a plate of food in front of her, the scent making her stomach growl expectantly. It was veal and roasted potatoes and carrots, each bite dripping with broth.

With her first bite, she was thrown back into her memories. In the safety of the Rosa, with few people around, she allowed herself to stay lost in the past, conjuring up tinier details so they would not get blurred by time. She clung to each of those details: a home with parents that loved her, the heavy blanket knitted together by grandmama, the warmth of cuddling to read a story together, the soft lyrics of a lullaby as she fell asleep.

∴ ∵ ∴

Theodora pressed her back against the closed door of her home. Quiet. No thrumming of a stelgladio and no commotion from her neighbors. She tugged the chain for her lamps, debating flopping into bed, but unsurprisingly, she found Maddox lounging in her chair.

"Okay, not only am I tired of seeing you every time I enter every dark room, but also you could have merely walked with me, and then you wouldn't have to brood in the dark by yourself."

She wanted to lay in bed, drink some more beer—courtesy of a Lawless she had found on her way back home—and not worry about her plans or agreements made at the castle right now. She wanted to worry about it in the morning.

"Well, you didn't come straight home, so I had no choice but to sit here. I didn't think going to the Digere was the best choice given the current situation."

"I didn't go there. I didn't want to risk seeing any more Satelles. I'm not that stupid."

Tension filled the space as their bantering died away. The weight in her gut built when her brain finally registered Maddox that was in her home. He'd been to the building before, sure, but never here—at least not to her knowledge. This was a place that was solely hers. Heat blossomed from her core, and she wasn't sure if it had anything to do with the two beers she'd already had.

"How did you get in here anyway?"

"Oh, you mean, how did someone break into a Lawless' home?" A smirk formed on his face—of course it did—before he continued. "Or how did a Lawless break into your home when you can't figure out how? Both would be terrible rumors to spread around Seclus and Lume."

"Seclus is a part of Lume."

"Is it?"

"What do you want, Maddox?" Theodora started removing her weapons and pouches, trying to ignore his snide remarks, before sitting on the edge of her bed.

"I wanted to talk about earlier tonight and to discuss our plan. I know the Satelles want you to meet the prince here in the capital house, but they made no mention of me. I'll admit they aren't the greatest at this plan-making business."

"Fates above! Can I say the fact that we host parties to celebrate upcoming festivals is absurd? I swear these royals have nothing to do with their lives."

"I have to agree with you on that. But with Lume's king doing everything, what is the poor prince to do?"

A chuckle escaped. As much as she tried to be serious and only business-like with him, he found ways to drop her guard. He made her forget the rest of the world existed. Just them. No titles, no Lawless, no politics.

"Seriously," Maddox continued, "wait until you see his shoes, and then you can tell me how bored he must be."

"His shoes?"

"You don't know about his shoes?"

"I tend to stick away from Satelles and princes."

Maddox raised an eyebrow. "Satelles? Really?"

"It was strictly business."

"For whom? You know Danika never saw it that way and I can't entirely say I blame her. You must be just as good at wooing as the prince."

Theodora rolled her eyes, "Well, back to the prince. What about his shoes? I feel like this is gossip that should be shared."

"No, you need to find out yourself. Consider it another small job."

"Speaking of... you came here with a plan, yes?"

"I'm going to ignore that question, Theodora. In all the fiedations you have known me, when have I never had a plan?"

"None yet. But I will say, when you don't have one, I'll be there to witness it."

"The Satelles mentioned you going, but they didn't specify why I am needed. Why hire both of us when this plan only requires you?"

Theodora leaned back on one hand and threw the other to her chest. "Why, Maddox, are you asking me out for a night in the capital?"

"Don't flatter yourself. We both can't go at the same time. It would definitely put the Satelles and their prince on edge. And we know the Lumens will recognize me as one of the Seclusians. Maybe you can use your...charm, as it were, to get a tour of the house. See what secrets might lay hidden there?"

"I see. Well, I think I can manage searching through the house during the party. I'm sure the prince will have a lot of people to mingle with. It still doesn't give you any involvement though."

"Yet. Let's see what you find, and I'm sure the need for my presence will become known."

"And on that, it's time for you to leave. It's been a day, and almost an entire night, and I need beauty sleep."

"And I don't?"

"You are not staying here. I don't know what this is," she said, gesturing between the two of them, as she rose from the bed to make sure he got the hint to leave, "so you're going to have to sleep in your own bed like a big little boy."

"Big little? You do realize that is a contradiction."

"Get out, Maddox."

With a smirk on his face, he rose from the chair, all regality following him. He somehow took control of every room with just his presence, and with his height before her now, he demanded her attention even more. Before stepping through the open door, he stopped himself. "Until next time, charm."

She watched as the door closed. When the latch fell into place, she released her breath. She made her way to it quickly, locking it. She'd thought Maddox leaving would help, but after that exchange, she was even more confused about where they stood.

Disobey

The metal of Rajveer's waist belt clinked as he worked to get it strapped across his body while swiftly descending the final stairs exiting the castle. The thumping of about a dozen Satelles' boots resonated around him, and he noticed Amicus, who immediately hurtled in his direction.

"No! Absolutely not." Amicus started pushing Rajveer backward, toward the doors.

"What are you doing?"

"No. What are *you* doing?" Amicus' voice dropped as he continued, "We set a plan in motion to have your father killed. I can't have you heading out and risking Lawless attacks."

"But I always go out with the Satelles."

"That was before this plan became imminent. With it already on course, it makes you the crown, and as captain, I'm not risking you for anything."

"It didn't matter before, Amicus."

"You're right. It didn't. Your father didn't care."

"I'm not some boy who needs to be protected. You've trained me as one of the Satelles. I'm just as equipped as any one of them." Rajveer couldn't stop the anger rising in his voice.

"Maybe so, but now you are important."

The words were a shock, as much of a shock as when his father choked him for trying to help Lume. Now Amicus cared?

"I'm sorry, I didn't mean it like that. Just… just stay here." Amicus placed his hand on Rajveer's shoulder, but it felt like a branding iron, burning through his jacket and deep into his skin. Rajveer jerked away.

But when his eyes met Amicus' again, there was no remorse. Amicus simply turned to his Satelles, and they mounted their horses and disappeared into the dark hours of morning.

Rajveer knew he should listen and follow his guard's order, but he was the fucking prince, and unfortunately for Amicus, Rajveer had spent all his bored hours training with his stelgladio to be a Satelle. He would be damned by the fates if he was going to sit idly by.

He made his way back into his room, locating an old hood from the depths of his disarrayed shelves. He tucked the fabric of the hood deep into the jacket before pulling it over his head. It wouldn't keep Amicus from noticing him, not for long, anyway. But he didn't need long, just long enough.

∴ ⁙ ∴

Rajveer had pushed his steed hard in order to catch up with Amicus and the Satelles. Fog swirled in the valley full of farm fields, granting him cover and allowing for his horse to blend in with the team of horses. The fog was so thick it caused moisture to form under his hood and along his hairline. A bead of sweat trailed down his spine beneath his shirt.

He and the Satelles raced along the darkened fields until a glimmering light appeared faintly in the sunken clouds. The small village of Parvos was engulfed in flames. The smell hit him first: smoke and burning wood, along with papers and fabrics and food and, he could only assume, bodies.

The Satelles scattered to investigate. The fire chased toward the walls, shadows mimicking where buildings had once stood. Rajveer dismounted and hesitated on the far reaches of the fire, searching the blaze for signs of anything.

Quick movement from his peripheral forced him to turn his attention back toward the nearest burning building. He squinted into the shadows, but between the fog and the smoke and the fire, he couldn't see a damn thing.

A bright blast of light hurtled in his direction, and on instinct Rajveer dodged, turning his back to the attack. The blast struck his cape, absorbing into the material. Shineblast. Well, at least he'd found the Lawless.

He tugged the cape forward with his hand, producing a shield across most of the front of his body as he attempted to track down the individual. Another blast. Rajveer shifted to ensure it struck the fabric. Now, he knew the direction it was coming from.

Rajveer rushed in the direction of the Lawless. Consecutive shots fired, each blast finding only his cape, absorbed and lost into the fabric. The Lawless, immersed in aiming with his shinegun, forgot to change tactics as Rajveer barreled in his direction. Rajveer smacked the weapon from the figure's hand.

The Lawless lost control of the shinegun. It flew away, sliding across the destroyed ground. Rajveer couldn't tell if the shape was of a man or woman, not that it mattered. A black hood and mask covered their entire face, and they moved in an inhuman way. Long sweeping arcs of limbs caught his attention before he realized a blade was coming next. Rajveer drew his stelgladio from its sheath, the shing of the metal sounding as he swiped it out before him.

The figure moved quickly, and Rajveer struggled to match their rhythm. He'd been taught swordplay as a dance, but all he felt now was the urgency of survival.

The Lawless' aggression with their blade was relentless, and Rajveer had nothing but instinct to defend himself, attack after attack.

Another swipe, but he wasn't quick enough. The blade cut across his shoulder before he could dodge it. The fabric of both his cape and tunic tore away, revealing a line of damaged, bleeding skin. Rajveer took a deep breath, trying to find a way to flip the bout around. He needed to take back the lead, be the one making the attacks, not wildly defending.

The Lawless pushed harder, pulling more strength into their swings, pushing more muscle into their blows. Rajveer stopped the blade once more with his, and the Lawless leaned into the standoff, attempting to overpower him. The Lawless used their other first to punch Rajveer's mouth, taking him completely by surprise. Rajveer stumbled back, the tip of his blade dragging in the dirt. He touched his mouth, pulling back to see blood coating his metallic fingertips.

When he glanced back up, the Lawless was arched back, bringing their blade down in a powerful blow. Rajveer forced his hand up. The blade struck Rajveer's metallic hand, and he grabbed onto it. He pulled with all his strength, forcing the Lawless forward in a stumble.

Dirt coated the black fabric of the Lawless' knees as he sank down. Rajveer debated removing the mask and hood to reveal their identity, but it hadn't mattered before, and it didn't matter now.

"Who sent you from Seclus?" Rajveer huffed out, throwing the Lawless' blade away from them. He slid his own blade back into its sheath.

"Don't know what you're talking about." The voice was deep and croaky.

"We know you are Seclusian, and we know Seclus is behind the attacks, so give me the name of the person giving these orders!"

The Lawless laughed. They were on their knees in the dirt of a field, buildings blazing around them, without a weapon, and yet they laughed—they fucking laughed.

Rajveer wanted answers, undisputed evidence he could give his father so the attacks could end. He stomped for the Lawless, grabbing them by the collar of their clothing.

"Give me a name!"

"A name won't help you, Lumen scum."

Rajveer punched the figure's collarbone with his metal hand. The crack confirmed he'd given the Lawless a fracture. In pain, the figure attempted to lean forward, but Rajveer pulled at their clothes, forcing them to look at him again.

"The name!" he demanded.

"What name do you want me to say? Altair? Vanmeir? None of those? What about Richard? Does that quench your thirst for vengeance, young lad?"

Rajveer shoved the Lawless away and rose, but they only laughed some more. The sound was loud, irritating and mocking. Rajveer turned his back to them, trying to gain his composure.

The Lawless' laughter ceased, and Rajveer drew his stelgladio at the same time they rushed for the discarded blade in the dirt. As they faced Rajveer, he shoved his blade forward quickly, forcing it into a space between the Lawless' chest and abdomen. The figure coughed, collapsing onto the weapon. Rajveer brought up his foot and kicked them off the stelgladio and onto the ground.

He listened to the sound of both the dying fire and Lawless. Rajveer heard shouts around him, but they were incoherent and muddled in his brain as his own deep breaths filled his ears. He turned, heading toward another house, to find more Lawless, when he stopped short.

Amicus stood wide-eyed nearby. Rajveer hoped Amicus wouldn't be able to figure out who he was. But the rage on Amicus' face was indication enough that his friend was far too clever. The smoke of the flames and the fog filled the space between them.

"What the fuck are you doing here, Raj?" Amicus looked to the motionless body on the ground.

"I told you, I won't just sit around. I'm not the bored prince everyone expects me to be."

"It isn't an expectation, Raj, and it isn't boredom, either. What happens when you become king? Are you going to continue galivanting off and risking your life so you aren't bored? I don't care about you being bored. I care about you being alive to rule Lume."

Other shadows formed, moving closer. Rajveer tightened his grip on his stelgladio hilt, preparing for the worst.

"Who goes there?" Amicus shouted into the fog.

Arms raised in complacency. "Just us, sir. We think we took care of all of them."

Amicus relaxed at the voice of one of his Satelles. "Any idea where these Lawless came from? Or what they wanted?"

Mumbles sounded from the other Satelles, but Rajveer already knew the answer. It was the same answers they had gotten after all prior attacks. They didn't know who in Seclus was behind them, or why they continued attacking random fields across the continent.

"Return back to the castle. Report any findings to your supervisors. And you, solider," Amicus continued, directing this order to Rajveer, "never disobey me again."

Rajveer debated letting the other Satelles know who he was, for them to know Amicus' orders only went so far against the prince. He wanted his Satelles to know that they would never fight a battle alone again, that although he was to be king soon, he wanted a different future for Lume. But he said nothing, instead biting the inside of his cheek before nodding in Amicus' direction. "Aye, sir."

Freedom

Sipping on a hot cup of yarrow tea, Theodora sat on the windowsill and watched the people go about their daily routines, the fog clouds following in their wake. Shadows of mothers carrying baskets and of carriages rolled along the cobblestones. Her favorite part was watching the children, especially the ones who were still young enough to keep their carefree view of the world.

She wanted to go for a run throughout the square to get some fresh air, knowing she'd neglected her training. She unfolded herself from the window and moved to leave, stopping before the side table where her dagger lay. Theodora should take it. Abandoning the bulk of her gear happened occasionally, but to go without a single weapon was a gamble. With a shake of her head, she turned away from the blade, bolted down the stairs, and headed into the square.

Without the physical weight of her dagger, she felt unrestricted, and it was intoxicating. A small taste of her soon-to-be freedom after she fulfilled this last task for Lume. Theodora weaved in and out of the crowds, the noise of the market square intensifying with each pound of her feet as she got closer. The smell of the smoked meats and savory breads made her stomach wake up, craving the possibilities. She debated stopping at Amabel's when commotion in one of the side alleys caught her attention.

Struggles in the alleyways weren't uncommon, with typical arguments happening over vendor prices, availability of items, owed debts, and men getting territorial over their partners. The fights moved to the side streets where the possibility of Satelles noticing them would be far slimmer. She tried to avoid them, but this one caught her attention, as even in the fog, the smooth movements of one of the individuals tugged on her peripheral.

She moved closer slowly to figure out why she was pulled in this direction. Approaching the group, her brain connected the recognition. Maddox battled five other men, his shinegun discarded uselessly on the cobblestones. Maddox tried to reach for his dagger, but the other men rushed him, battering their fists into his body. The grunts were a symphony, a crescendo of noise the longer the fighting continued. Theodora noticed they were toying with him, dragging out the fight, attempting to wear Maddox down.

Entering the shadowed path, Theodora darted for Maddox's shinegun. She cursed the fates and Seclus' device registration rules. She needed to get it altered quickly if she had any hopes to use it. The coarse rock of the ground cut against her knuckles as she picked it up. The stones showed it still had charge, and with a small brush of her thumb on the trigger, the safety light blinked blue.

Apparently, Maddox's shinegun was registered to work under her prints, but she didn't have time to analyze the reasons. She removed the safety, turning to one of the men standing further back from the group. She hoped that with one blast she could surprise them enough to force a retreat. She brought up the shinegun, the weight slightly off compared to her own, took aim at the man's back, and fired. The lighted charge sailed across the small space, bursting into his skin. He screamed in pain, trying to contort his arms to hold the wound now flooding the threads of his tunic with blood.

Maddox spun his attention in her direction, and she didn't miss the shock that ran over his face when his eyes found her. Every one of them now turned to face her. She aimed the shinegun at the next man, waiting for the group to decide. Three of them turned and bolted, but the final one remained, turning his body to Maddox instead of following his companions. A flash of silver caught her attention. Without logic, she flung the shinegun to the ground and raced to them, attempting to stop his arm from reaching its target.

But the man was too close to Maddox. Her response had been inconsistent with her expectations. Theodora hadn't reacted reasonably when she threw the weapon to the ground, racing against time instead of remaining efficient. The man buried his dagger deep into Maddox's ribs, the blossoming of bright crimson jarring against his white tunic. Maddox staggered back, gripping the building face before he slid down to the ground. The man abandoned his dagger submerged in Maddox's side and fled as Theodora kneeled, placing pressure on either side of the blade.

The warm blood coated her palms quickly. Her mind raced with her options, but leaving him here, stranded alone, wasn't one of them. The sensibleness of finding help left her in a chaos of emotions as she looked down the alleyway toward the main road. She needed someone, anyone, to show up. A few people strolled by, but she couldn't find her voice, the words stuck in her throat beneath the knot that had lodged itself there. Another round of people entered her line of sight, and she choked down another gasp. A man stopped sweeping the alley and said the words she had been unable to speak.

"Over here," he shouted, and two more shapes formed in the fog behind him, moving closer. The man made it to them first, immediately dropping down to tear the fabric of Maddox's tunic, investigating the wound. As he did his own search, Theodora did hers on these three individuals. She noticed that all three must have had some type of training, as she could see the definition of their muscles even through their clothing, but she didn't recognize any of them. The man had dark hair, a similar shade and style of Maddox, but his eyes were green.

She remained crouched near Maddox as she took in the appearances of the two women next. They were clearly twins, with identical builds and features. They both had hundreds of tiny braids crowning their heads with bright pink and purple braids peeking through a sea of night. A sense of still calm radiated from them, which was unsettling given Theodora's panicked mistakes.

The man rose, searching the main street behind them. Theodora stood up as well, focusing on the noises for any indication of Satelles. Nothing, yet.

The man lowered his voice. "We've gotta move before Satelles get word and investigate. The dagger looks to have missed anything important." Maddox interrupted with a slight groan as he attempted to shift against the wall. "We need to get the blade removed and get him to a bed quickly, so we can have it searched more thoroughly and cleaned. I don't think the blade was poisoned, but it could be slow acting, which means being in a better location than this would be, well, better."

Maddox remained slumped against the wall. Theodora bent down. His eyes remained closed and he breathed deeply as she swiped a strand of hair away from his face, memorizing the way his skin crinkled around his eyes. The two women inched closer, and Theodora moved her body in front of him defensively.

"What do you think you're doing?" Theodora wanted to fill the words with more malice, but instead they came out wounded.

The twins' only response was moving their hands to the weapons at their hips.

It was the man who spoke. "Theodora." He dipped his head. "We are Maddox's—entourage, as it were. We are going to help him and you're going to go after the ones who did this."

Maddox had friends? She had never seen these three in her life, but as she stared at them and their calming presence, something deep in her gut told her they would keep Maddox alive, at least long enough for her to track down the ones who'd done this. She moved to get up when a hand grasped her wrist. She looked down, seeing Maddox's fingers wrapped firmly around her arm, the warmth of his hand seeping into her skin.

"I'm going to go after the people who did this, Maddox."

A sad smile cracked his face. "Be careful. And remember, you're supposed to be Lume's hero after all this."

The man knelt and prepared to take the blade out, so she talked, hoping it would distract him. "I'm not, Maddox. When this is over, I'll be an outcast and spending the last of my fiedations alone and away from all this."

"You're wrong. You never see the big picture, charm. At the end, you will have saved Lume and—"

"Go!" The man interrupted them.

She pulled away quickly, noticing how Maddox's jaw clenched in pain. She took a slow step back, distancing herself from them, from him. When faced with the possibility of last chances, emotions dragged down her protection, ripping apart the walls she'd placed around her heart. The possibility of future regret lingered in her mind. She debated kissing him. The thought was conjured hastily, but quickly became an ever-consuming thought. Except maybe this thought wasn't hasty, maybe this was something far deeper she had never acknowledged before. Whatever it was, she refused for this to be a goodbye, refused for this to be the last time, the last chance, the last moment... the last anything.

The same smile he had given her before now reflected on her face. Before turning to hunt down the attackers, she whispered to no one, "I was never meant to be a hero."

Portal

The pain in Maddox's ribs was like a fire he couldn't put out, and damn it all if Valix didn't keep adding fuel to it by pushing on the wound, prodding the blade. "Are you trying to kill me?" Maddox wheezed out.

"Don't be dramatic. You've seen worse wounds than this."

"How long are we going to sit here?"

"The fog makes it more difficult. The Gems are checking the street and then we should be good."

Maddox's neck was growing stiff from the way his body had slouched to the ground. "Can I at least stand up?"

Valix looked at him like he was torn between rolling his eyes and smacking him, but chose neither. Valix knelt lower, tucking his arm under Maddox's back to help him rise. By the time Maddox had himself steadied, the twins were returning.

"After the carriage passes, we are clear," Jemmie said without a backwards glance.

The sounds of the market fluttered down the alley, and the persistent squeal of the hulking carriage grew and faded away. Maddox waited a moment longer before struggling to tug the pocket watch from his vest.

He pushed down on the turner, allowing the metal to spring away from the face of the watch. He spun the metallic disc that rested on the face, counting the spins. Three clockwise. Eight counterclockwise. He pushed down the turner. With a swipe of his thumb, the face rotated out of the traditional setting, opening a circular portal within the middle of the circle formed by their bodies.

They stepped through, the electricity rippling around them as they left the alleyway and entered Maddox's room at Previt.

Unknown

The trek to Seclus had felt longer, but time slowed even further as she sat in the corner of Ludi Votivi. Theodora had learned the five men were Seclusian—or now four, since she'd found the one whom she'd shot, dead on the side of another alley. Their darker clothing and thick fabric were indicative of their origins, since the topside weather was as indecisive as a child. She had also rummaged through the dead man's pockets and found a gambling piece from the Ludi Votivi den, erasing any further doubt.

The barkeep, who she learned was named Gustavo, continued to keep one eye on her. Few people had been in for their drinks, served by the waitress, Camilla. Her experience here was evident in the ease with which she remembered customers' names and the small facts they exchanged over their beers, covertly persuading patrons to buy more rounds of drinks.

Theodora had planted herself at a table in the back of the room, her beer mostly untouched, taking in the view of assorted tables. Some were empty planks for food and drinks, while others were circular and used for local games, with smooth trays carved along the edges housing gambling chips.

The morning fight replayed in Theodora's mind like unending torture as she noticed every mistake she had made. Why had she abandoned the shinegun and run for the dagger as if she would be fast enough to stop it?

Theodora didn't know where exactly the attackers had run off, and if they'd escaped the capital completely, her revenge would be forced to follow her like a vulture, waiting for an opportunity in the future. But she hoped they would be proud of taking down Maddox, so smug and high off their valiant efforts that they would stay and brag to their local compatriots.

Maddox. His name acted like a dagger of its own, slicing away at her heart. Vulnerable was a word Theodora didn't want to be in this world, or probably any world. She'd hardened her heart against society, built a reputation that secured her life and klaud, and had been searching for her opportunity to flee.

Danika tried to present one to her in a marriage offer, which Theodora had strongly considered. Danika, impatient to fix her and to erase her past and Lawless ways, became desperate. Eagerly, she'd told Theodora daily of her promises to get her crimes pardoned and remove her from Seclus. But no matter the promises Danika made, none of them could remove the blood staining her hands.

And then there was Maddox, her hypothetical in the world. They were two sides of the same coin. They had watched each other barter with lives and wreck their souls to obtain things they needed. But two sides of the same coin could never come face to face. She never let her mind wander past anything more than being associated with him. With Danika, the emotions never really existed, at least not for Theodora. With Maddox? She drowned in emotions. She kept them bottled deep within, refusing to allow them to open to the world. Witnessing his bleeding body in the alley had been the final blow, shattering the bottle, and the emotions she spent so long containing had released with no possibility of return.

Theodora was breaking their unspoken agreement by protecting him. Avenging him. Then again, she already had when she'd agreed to work with him.

Maddox would be her undoing.

The door banged open, the scene disappearing from her mind's eye, and the first of four men practically fell into the den. *The* four men. She watched as they ordered drinks; however, their demeanor indicated these weren't their first beers of the day. Camilla sauntered over, her dark braids swaying with her movements as she delivered the round of beer. The men's' loud exchange about the alleyway fight boasted throughout the room as they told everyone how they had stabbed a man. Theodora noted, however, that they never said his name.

Weaving through the tables and a handful of other patrons who had walked in, Theodora stopped directly in front of the four. The first, Lip Mole, had pulled his mug upward. "Are you here to show me a good time, honey?" The words were barely separated, a slurring of constants and vowels, punctuated by a ridiculous laugh. The others, who she had named in her head as Black Hair, Missing Tooth, and Nose Ring, joined in his humor. Clearly their prior drinking had erased their recognition of her.

Camilla approached again, a tight smile across her thin lips. "Would you like me to move your drink over here, miss?"

The few patrons of the den had perked up. This is what made Seclus incompatible with Lume. Everything, down to the basic social rules, were different. Fake justice was not beaten into you by Satelles; instead, swift retribution was accepted by fellow Seclusians.

"No, thank you." Theodora replied. "This will only take a moment."

Camilla turned, but her attention remained on them.

Theodora leaned forward, grasping the table with both hands, a pout forming on her lips. "Such a shame that you don't remember me." Confusion plastered across their faces, and they exchanged blank stares. With a sigh, she stood back up, crossing her arms over her chest, and offered, "I'm the one who found you beating up Maddox."

There was still a delay, but soon the realization became clear in their eyes. Theodora moved. She grabbed the closest on her left, Black Hair, by the collar of his tunic and pulled, her right fist striking his nose, awarding her with a satisfying crunch as blood sprayed. He yowled in agony as the other three began to shift as well.

With her hand still fisted in the first man's tunic, she reached around and pulled the dagger from his hip. She threw it at Lip Mole, who was scattering out of his chair. The blade met its target, the meaty flesh between the man's chest and shoulder. It stopped his progress as she punched Black Hair to the ground. Missing Tooth and Nose Ring shoved through the doorway into Seclus, slipping over each other as they bolted in opposite directions. She let Nose Ring, the one who had stabbed Maddox, go as he headed in the direction of the tunnel entrance.

Tracking Missing Tooth, she pulled Maddox's shinegun from her waistband. She looked past the discs on the top of the barrel and watched as he raced across the elevated walkways, bolting up various sets of stairs, moving in any direction to expand the gap between them. Taking a moment, she breathed out, let her mind make the calculations, and pulled the trigger. The charge shot toward his back and dropped him to the ground.

The witnesses watched him for a moment before turning back to their work. Good old Seclus.

She didn't wait a second longer before bolting for the tunnel entrance after Nose Ring. She shoved past people in walkways and raced for the market, hoping the drinks he'd consumed would slow him down. She figured Black Hair and Lip Mole were still alive, and Missing Tooth was questionable, but she didn't care. Her revenge waited with Nose Ring. She spotted the whisper of his shirt in the crowds, but instead of following his path, she ducked into a different alley, running along the backside of the buildings. She took advantage of the fact that Nose Ring was fueled on beer, adrenaline, and the need for survival. Theodora was fueled on vengeance.

Tucked in the shadows at the corner of the alley behind the main street's row of shops, Theodora listened. Few people used these alleys, except for workers who dumped buckets of water. Nose Ring's wobbled footsteps slapped as they approached, punctuated by the occasional splash as he found the puddles riddling his path.

As she peeked around the corner, she saw him. He had turned his head to look behind him, prey attempting to find the lost predator. When he faced forward again, Theodora stepped into view. He yelled, his voice responding well before his body could. His feet jerked to a stop, causing him to slip and fall back, landing on his right hip. He groaned and tried to pull himself up, dirt and mud sticking to his clothes.

Theodora moved closer and placed her boot on his ankle, pressing her weight into him. He stopped his progression upward as he reached for the source of his pain.

It felt good to her, the feeling of his bone against the bottom of her sole. He laid back, pain still contorting his face as he panted. Fear shone brightly in his eyes.

"Who sent you to Maddox?"

"Wha—wha—what?"

"Who sent you to Maddox?"

"I don't—I don't know what you're talking about."

"I don't have time for games. No one, not even five of you, would be stupid enough to go up against him alone in an alley without a good reason. Who sent you?"

"Honestly—" He continued stammering, searching his brain for words.

She raised her hand holding the shinegun. She admired the look of the design and the amount of charge it still held, while she gave Nose Ring a chance to figure out how to use his mouth. She aimed the barrel at his head, making it clear when his opportunity was gone.

"Please," he began, fear dripping in his golden eyes.

But reasoning was something Theodora couldn't process right now. The man in front of her may have been drowning in beer, but she was drowning too, in a sea of emotions. A sea of the unknown.

She didn't know if Maddox was going to live. She didn't know what this meant for their barely laid loose plans to assassinate the king. She didn't know what Seclus or Lume would do if they aligned themselves. But with all that she didn't know, she knew Maddox was worth it. With Theodora's attacks on these Lawless, it showed the world that he meant something to her.

She braced herself for the consequences, the ghosts that she would have to live with. She took a deep breath in, and as she breathed out, she pulled the trigger, and the charge ignited his skull.

∴ ∵ ∴

Theodora didn't know the name of the man she killed the day before. No matter how many times she pulled triggers, it didn't make it any easier, and although she had taken revenge, she felt hollow. Theodora had never been told where to find Maddox. She waited, waited for him to appear like he always did, to find out if he was even still alive.

She had laid in her bed for over a day. She watched as Fiedel changed colors through the curtain fabric, watched as the shadows danced around the room with the passing of time. But Maddox, or any of his so-called entourage, never showed.

Rajveer's house party loomed over her. Whether Maddox was alive or not, assassinating the king was the only way out of this city. With a groan, she dragged herself from the bed—okay, she fell out of bed. She hardly tasted the strip of dried pork and fruit she chewed as she tucked her ivory tunic into her pinstripe trousers. Strapping suspenders over her shoulders, she slid a small dagger into a discreet pocket sewn into the leather of the straps. Prince or not, she wouldn't make the same mistakes twice.

As she stepped outside, the warm air wrapped around her, Solta holding onto the world as dusk barely held onto the sky.

Regret

Rajveer mounted his steed, settling himself into the saddle as Miles climbed into his own. Although Fiedel had started its descent, the Solta heat clung to their world. It made him feel frazzled, especially after his explosive dinner with his father and the fact that he'd pissed off Amicus. He was grateful to not have to wear the Satelle golden jacket and cape Miles had donned.

They let their horses meander leisurely through the trees, following the path from the castle into the capital. The scent of honeysuckles and a hint of apple drifted through the humid air. Rajveer felt tension between himself and Miles, like a physical weight was pressed on his shoulders.

Rajveer tried to focus on the chirp of the birds and the slight thud of the horses' hooves on the dirt path, but the damn uncomfortable glances got the best of him.

"Fates, above! What is it, Miles? Out with it already!"

Miles abruptly turned and faced his prince, shocked at the accusation, as if Miles thought he'd been subtle.

"Oh, don't act all surprised," Rajveer went on. "You are about as obvious as fire from a foco."

"Sorry, sir. I didn't mean any disrespect…"

"I don't care about respect. Tell me what's on your mind."

The silence tumbled between them. Rajveer didn't want to push the young Satelle, not with everything he had already gone through. But how could he trust a soldier who refused to share secrets with him, who failed to voice his grievances and opinions? Rajveer vowed secretly to himself he wouldn't ask again.

The conflict on Miles' face meant he at least debated speaking what was on his mind. Long, desperate minutes passed, and Rajveer didn't know if he would be able to stand by his own word. Miles' meek voice finally broke forth, barely audible over the birds.

"Why do you want to kill the king? I mean, he is your father."

Rajveer's mind whirled through an inner debate. He sought the words, the proper explanation. Miles, orphaned as a baby, had been raised by a Lumen family until he was of age to become a Satelle. He wouldn't understand what it was like, to have a father who barely even earned the title. King Klauduisz had helped Lorelei conceive him, but did that act make him a father? To Rajveer, the king was but a man, a man that had stood in his path, refusing to let Lume become the grand capital it was meant to be. A capital of strength and resources. And pride.

No, the man that called him son didn't know what pride meant anymore.

Miles' doe-like eyes dug into his own. Rajveer turned his attention downward to his own lap, his metallic hand wrapped around the reins, and a past life of tainted memories seeped forward.

Rajveer thought he had been loved by the king. As a young boy, he dutifully followed his father around the castle to meetings with Satelles, court with lords, and, occasionally, training. As he got older, he traveled with the royal procession to visit the neighboring towns, more times than not getting treated to a peppermint ice stick.

What he hadn't realized as a kid was that it was all merely a plot for him to learn his role, to be molded and shaped into what his father wanted and expected. It became apparent when his relationship with.... No, he was far too sober for such thoughts.

"Miles." Rajveer's voice matched the Satelle's prior weak voice, as if he was afraid the world might finally hear the truth. "I can never answer that in such a way you will understand."

"Can you try?"

A deep breath filled his lungs as he attempted to soothe the emotions that burned now. "He is just a man, Miles, a man who may have fathered me, but has not been a father since long before the queen's passing. And he... he doesn't see Lume's potential. I want Lume to take back the power it has always had, show Seclus and any of those that threaten us that we won't just pass them our crown. And he *had* listened for a time and trusted my input, especially with Alouette around. I am sure you have heard the rumors from the people."

"I have heard some, but even Amicus wouldn't go into detail to tell me if what I heard was true."

Damn Amicus for keeping his mouth shut. It would have spared him now. "They probably were true. I loved her. And both the queen and king loved her. We were meant to be wed. But on the night before the wedding..."

He couldn't continue, and luckily Miles didn't push. It was ultimately Rajveer's parents' fault he ended up like this: drinking into oblivion, losing himself night after night. All because of Alouette leaving in the cover of darkness to travel back to Freta. He had loved her. He had loved her fiercely. "I thought she was happy. *I* was happy. The Lumens were fanatical about her. She was majestic and kind, like she was born to be queen."

"I remember seeing her in the capital when I was a child. She was beautiful."

"She was." A weak smile appeared as Rajveer thought of her. Of the memories. "The king had made her duties clear to her, what he expected of her when she became queen. And she had never backed away. But after I proposed and the wedding details were being planned, the king made his true wishes known. He made it clear that after our wedding, we were going to be crowned as king and queen. He wanted to abdicate the throne and live out his older days with his wife. He said he wanted to witness me rule and wanted to watch his grandchildren grow up.

"It was too much for the wildness that lived in Alouette's heart. We had grand plans of what we were going to do during our time as merely *heirs* to the throne. She wanted to travel to all our cities to see what they were like. Meet the people. Talk to our future citizens. Taste the world. She wanted to help me create a better system than what he had. One that helped unite our peoples together, remove the wedge and start a council of decision-makers. But the king choosing to step down immediately wiped away those plans.

"She feared that with him around to *watch* us rule, he would ensure no massive changes would be made. I would have no legacy to my name. Just a prince-turned-king too early in life, leaving my name to one day be forgotten to time. So, she fled. As far as I know, she is still out there, living our adventures."

Miles was quiet. Rajveer didn't know whether he had given Miles the answer he sought.

"Do I want to kill him?" Rajveer continued. "No. But he has told me numerous times that the only way he will abdicate is by me marrying, and as you can see, he destroyed my best chance at that. He is a poison to Lume, one that needs an antidote before it becomes too lethal."

"Do you not want to get married?"

"I don't see how that is possible anymore. My heart is with Alouette in Freta, and most Lumens I have attempted to court are either so in love with the royal life that they are blind to the idea of change, or are fearful of the idea of becoming any sort of ruling partner. It isn't for lack of trying that I haven't found someone."

"Well, you never know, Rajveer; you are meeting someone new tonight."

He chuckled. "Yes, a Lawless. I doubt she has any desire for royal life."

"But she might not be blind to change."

This settled into Rajveer's mind. Had he thought about this the wrong way? Could he use Theodora as his way to the crown, not through killing his father but by marrying her?

They exited the woods into the outskirts of the capital. On the streets, more Satelles joined up with them, their team of horses forced to weave through Lumens. Rajveer wanted to speak again, wanted to talk this out with Miles, but he couldn't deny his people. They grabbed his attention whenever he was around them. He nodded to the men in acknowledgement, and waved to the children that hid behind their mother's wide skirts. The women's smiles were wide and hopeful.

His people brought such a thrill of excitement to him. It made him proud to be their prince, but he wanted to be more, needed to be more to help save them from their destitution.

They traveled along the street of Prae to the royal family's capital house, knowing the staff had spent the day preparing for this party. Rajveer pushed aside the dreaded, morose conversation he and Miles had exchanged and forced himself to drown in the Lumen excitement.

As they continued forward, he realized he was going to finally meet Theodora. But after these fanciful ideas had bounced around in his head, he wasn't sure he was ready.

Without a second thought, he jumped down from his horse and walked forward with the reins in hand. The rest of the Satelles remained on theirs mounts, keeping their attention on his safety. More people flocked to them, slowing down their pace as he kissed the brows of children, grasped the hands of men and women. He was prince, after all, so he could certainly be late to his own party.

Witness

Positioned a distance off the main road was the capital home, on the outskirts of the market. It was a grand red brick building which jutted up from the flat surrounding yard. As Fiedel fell beyond the horizon, a pale violet painted the sky. Blinking bugs scattered over the grass, and insects roared to life. The first floor of the home was lit and lively, curtains drawn back, allowing glimpses of the shadows from the people within. The smell of food wafted out of the windows, which had been cast open, hoping for a chilled breeze this evening. The rumbling of people and boisterous laughs floated down the pathway to greet her.

Theodora's stomach turned, at first in delight at the thought of food, but then souring because of the copious amounts of people she was about to face. This was much more than any task she had ever accomplished before. This was an act, a façade for all Lumens.

She kicked her feet along the path, ignoring the dust and bits of gravel. Couples walked past her, leaving the party already, tight nods and pulled smiles across their faces. Her thoughts betrayed her as they fumbled to Maddox. She shook them away like an irritant, reminding herself to stay focused on the task.

With the front door open, Theodora stepped over the threshold, sweeping her gaze over the vestibule, and stopping when she found an icy stare from one of the staff. The woman bustled forward through the bodies, her golden-colored ringlets bouncing. Although she nodded her head in acknowledgement,

Theodora still caught the disgust that slid across her face as she took in her attire.

"Good evening, missus. Are you here to deliver more food?"

"No, I'm here for the party."

The woman looked around her, searching for something. "Alone?"

"Not for long," Theodora produced with a wink.

The woman paused again, and Theodora noted the dispute in her eyes before she stepped out of the way and gestured into the home.

Before Theodora headed into the depths of the party, she caught the staff member by the shoulder as she was getting ready to greet another group. "Where can I find Prince Rajveer?"

"Prince Rajveer has many requirements in his life. We can only hope he can attend." She pulled from Theodora's grasp and turned back to her duties.

Entering the receiving room, she found it impressive, although it was probably what was expected for a royal home. Here the walls were smooth and neutral-colored, almost like cream had been splashed onto them. They were trimmed and molded with a warm walnut, which spilled down on the floors that were glossed to a bright shine reflecting the lampposts.

Along the largest wall artwork hung, showing murals of the three largest cities on the continent, outside of the capital itself. The docks of Freta extending like fingers into the choppy teal waves, vicious and menacing as they surrounded the port city. Vindem decorated with brilliant grassy hills, grape trees and vines snaking around the wine city and the border of the art piece. The last painting was filled with large red trees and the mysterious lighting of Conlis, the city buildings tucked and hidden in the encircling ancient flora.

Theodora chuckled to herself, thinking of adding one of Seclus, its recent impact on the capital growing as it produced similar resources as its counterparts. But to imagine a watercolor painting of it? It would be dark and ominous against the wall, no color capable of capturing the real essence of being underground. That essence wasn't the stacked buildings formed into cavern walls or walkways intertwined in every direction. Fates, it wasn't even the chandelier. It was the sounds: the echoes of every boot, every strike of metal, every punch or blast...

Laughter cut into her thoughts, and Theodora turned her attention back to the receiving room. Settees of varying sizes in beige and brown damask were placed in various arrangements. A large archway allowed a glimpse into the dining room with a massive table filled with different foods: roasted chicken, tenderloins, cheeses, fruits, potatoes topped with bacon, roasted vegetables, and sweets sprinkled throughout.

She felt small in comparison to the amount of luxury and grandness before her. The rooms were filled with tight coats, button-down shirts, cravats and cufflinks, and hoop skirts with warm fabrics the colors of astrum: golds, rust, butterscotch, and pumpkin. Pulled corsets and vicious necklines flaunted gaudy jewels. Theodora looked down at her own attire before inwardly chiding herself for not remembering Lumens didn't share the styles of Seclus.

Moving to grab a small plate of cheeses and nuts, she scanned her eyes over the swarms of people as they hovered together, chattering mindlessly. Staff weaved in and out, taking empty tableware and flaunting full glasses. Theodora meandered through different groups, surprising most with her attendance. She always pushed off their inquiries indicating she *had* to go to a capital party at least once in her life. No one could refute that.

She spoke to those she recognized from the market, and others whom she'd never spoken to before would usually chime into the conversation. No, she didn't know Cassia's mother had passed away. Otto did deserve escaping the market to enjoy his old age out in the woods. Yes, she did know that Kaamil was with child again. Luzmik needed to become wiser about his klaud or he was bound to lose it all. She hadn't seen Earleen since he had fallen and hurt himself.

She popped the last bit of cheese into her mouth, and immediately a staff member took her plate away, replacing it with a fluted glass of wine. It was a rarity the citizens of Lume were only given when the king allowed it, which explained the masses of people who attended even for a brief period.

Sipping on the drink, she noticed Danika enter the room. Theodora refused to give her a smile, the emotions still too raw. As Danika slipped into the group discussion, the others immediately turned to give their good graces to the esteemed Satelle. While Danika answered their questions, her gaze reverted to Theodora's often, an unspoken question lingering. As the conversation changed, Theodora turned to leave with Danika quick to follow suit, placing her hand on the small of Theodora's back, guiding her between clusters of people, back to the entering room. Danika pushed her into the corner, blocking her view with the closeness of her body.

"Where have you been, Theo?"

"What do you mean?"

"I've heard the gossip, that you went into Seclus causing a huge commotion, using flash crystals, and killing people. That you had this huge scheme with Maddox? This really isn't the time for a rebellious act from you two, not when there are other plans in motion."

"Seriously, Danika? I didn't use a flash crystal. And I only killed one person, I think. Maddox killed the other. And why are you listening to capital gossip?"

"Why would you start teaming up with Maddox when we already have this other—thing." Her voice dropped to a whisper as she scanned the people around them to be sure no one was listening.

"I'm sorry, but didn't you and everyone else ask Maddox and I to work together?"

"Not like this. You're supposed to be staying low. Wait for the prince to meet you and ask you to the festival so that we can present you to the king."

"Wait, Rajveer is going to ask me to the festival?"

"I—I wasn't supposed to say that. He wanted to surprise you with it."

Theodora had calmed herself knowing the rest of the capital would be aware of her going to the castle under the ruse of being Rajveer's *entertainment* for the night. But now he was planning to make it seem like he was *actually* courting her? After she had started making her feelings known to Maddox? Nothing good was going to come of this, and her gut tightened.

"Do you know if Rajveer is planning to attend tonight? Or was that all another ploy?"

"He should be, but he is also known for doing whatever he pleases." Danika continued, but Theodora failed to hear the rest. She rolled her eyes and turned back to the crowds of people, glancing quickly over faces to see if anyone could provide insight on this evasive prince, when she found one. The mumbles of voices drowned away as she focused on the man who entered the vestibule. The black of his trousers, the black vest swirled with a rich mahogany color, and a crisp white shirt blended into the clothes of the other partygoers. His black hair shimmered in the light as he turned away into the crowd, the cravat around his neck like a beacon.

She lightly touched Danika's arm, cutting her mindless chattering short to follow the ghost of a man. Danika scanned the crowds, trying to discern what had grabbed Theodora's attention so strongly, but everything stopped short. The dull roar of voices quieted as everyone turned in the direction of the front door, forcing Theodora to stop her chase and follow the crowd's attention. Hoop skirts pushed and cramped against one another, creating miniature blockades. Others leaned into each other, whispering secrets in their anticipation.

The woman who had admitted Theodora into the house appeared first, gracefully clearing her throat with a curtsy. "Prince Rajveer, heir to Richard Klauduisz, King of Lume."

The prince and a Satelle stepped into the room as the Lumen people moved into their usual motion of respect, forming their right hands into fists above their hearts and bowing their heads to the floor. She felt Danika doing the same next to her. As everyone fell into their regards for the prince, Theodora remained straight-backed and unmoving, watching this man across the room. The murmurs of "long live the king" and "fates bless the king" overlapped quietly in a wave through the people.

She'd heard much about the infamously spoiled prince from the whispers and talk among the Lumens and Seclusians alike. Although she had seen him from afar, she typically dodged any type of event or slipped off before she could really come face-to-face with him. At first, it had been due to fear that she would be recognized as the girl who had fled from the king so many fiedations ago, but as time warped the past and she grew into a different person, she still never seized an opportunity to meet him.

Theodora thought he looked younger than she'd expected. He was slender and had some height to him, but his trained muscles couldn't be hidden beneath the light robin-egg colored shirt he wore, which he'd accented with a long tie of navy bearing swirls of matching tones.

The prince's eyes caught hers through the bowed Lumens before his modulated voice filled the room. "As you were, my fine people. Tonight, we celebrate the coming of the Astrum Festival. Drink your wine, enjoy the great food, and be merry with whomever you favor."

For a prince with little power and nothing to do, at least he was charismatic.

The people turned back to their groups, the murmurs intensifying and laughs filling the space again. Other Satelles entered the room, taking places around the perimeter of the home as Rajveer immediately approached her.

Theodora stared into his eyes. They were deep cobalt, like how she would imagine the sea would look at unimagined depths. They were bright and alluring. His golden blond hair was shaped upward down the center and sleek on the sides, streaked with various shades of color, as if kissed by Fiedel, that filled the place of his missing crown.

He was regal and powerful, pristine, with not a piece out of place. Or so she thought until she caught sight of his left hand, which he kept down near his body. She honed in on it. It appeared like his entire left hand was missing, and in its place was a skeleton-like metal replacement with gears adorning the knuckles. It was a dark metal that emitted a smoothness. Something of this magnitude, especially if it was functioning, was inconceivable. Whomever had crafted this was someone she needed to find. Why had all the town gossip never whispered of this?

She stared down the heir to the throne as he stopped mere inches from her. He didn't tower over her, but it did force her to raise her head to meet his gaze. He swiped a glass of wine from one of the staff and raised it to his lips as he spoke.

"I've never seen you before."

"I never wanted to be seen, but I decided I needed to witness for myself this beautiful prince everyone was talking about."

"And? Did I disappoint?"

She glanced up and down his body, mimicking his assessment of her. "I'm not sure yet."

He reached his skeletal hand to her face, letting a fingertip slide down her cheek. His eyes were as piercing as the cold metal of his finger, which spread from where it met her skin. "Well, I guess I will have to persuade you. Walk with me."

He dropped his hand and tucked it into a pocket while extending his right arm to her. She looped hers into it, and they made their rounds about the lower level of the home, mostly speaking to people Theodora had never met before. Almost all of them hid repulsive looks of her as they approached. Nonetheless, she couldn't deny how much the Lumens adored their prince.

The forced courtesies she was exchanging were growing tiresome and wearing on her when a hand grazed her shoulder. She turned, expecting the dark-haired hallucination she had seen earlier, but found Amabel instead. She jerked away from Rajveer to pull her into a hug.

"What are you doing here?" Amabel insisted.

Clearly the question of the night.

"I decided it was time to finally see the prince up close."

"You should have told me you were coming. We could have talked about attire."

"Why? And look like all of you?"

"I don't know. I think you would look ravishing in a lavender dress."

"I think she already looks ravishing." Rajveer stepped forward, gliding his hand around Theodora's waist.

"My prince." Amabel placed her hand over her heart and bowed deeply. It was odd to see someone Theodora knew so well bow to her, even though she knew it was really for the man next to her. But the rush of power and possibility wasn't lost on her either.

When Amabel rose, she struggled to find words again, uncertainty descending on her face.

"Prince Rajveer, let me introduce you to my friend, Amabel. She makes the most amazing chocolates." Theodora raised a hand in Amabel's direction.

"Oh! Of Amabel's Sweets? I love the chocolates filled with the popping rocks."

"Popping rocks?" Theodora whipped her head in his direction, for she had been a fan of Amabel's sweets for a time, but never saw popping rocks chocolate before.

"You haven't had one? When you bite into them, they are filled with a sweet nectar, but also cause this popping sensation, like flares in your mouth."

"You are too kind, my prince." Amabel's blush crawled down her cheeks and her neck.

"I can't believe I haven't had one before." Theodora attempted to fight back the shock of never seeing them in all her fiedations of being a patron.

Rajveer cleared his throat, pulling on his cravat. "Amabel only creates them for the festivals. Maybe you can attend the Astrum Festival with me, and I can witness your first taste of them."

The circle of those around them dropped their conversations, their eavesdropping clearly apparent. A small silence fell amongst everyone nearby, as if a collective breath was being held.

Theodora bowed her head. "It would be an honor, Prince Rajveer."

"Excellent. Have you tried her chocolate-covered pineapples?

Theodora moved to respond, but a Satelle walked forward, interrupting them. Theodora watched as he whispered into Rajveer's ear. "I'm sorry to leave you, but I am needed for a moment. Do let us catch back up later."

Rajveer bowed to Amabel before pulling Theodora's hand to his lips. The kiss was quick as he kept his gaze firmly on hers. She knew this was part of the plan, but she couldn't stop the flush of her cheeks in knowing everyone was watching them at this moment. He turned away from them to leave for whatever forced him away, and Theodora was grateful for the opportunity to confirm the phantom she'd witnessed earlier.

Promise

The capital house study featured wood floors and cream walls that continued from downstairs. When Maddox entered the room, he noticed the left side was an alcove of shelves full of books with an array of darkened and worn spines. They appeared to be mostly ledgers with a handful of novels thrown onto the top shelves. In the middle of the room was a wooden desk made with the same wood as the floor, making it appear as if it had grown right there in the room.

Maddox walked along the shelves, perusing spine titles for anything that was of intrigue or purpose. He had stolen up here upon the prince's arrival. He may have told Theodora to come to the party alone, but it had never been his intention. He wasn't willing to allow the pieces to move erratically along the board.

He moved to the desk and fiddled with the papers there, the smell of old pages floating up to him. He noted the date on some of the top sheets. They were from ten fiedations ago, and if they held secrets before, they had long since become stale.

Not ready to go back down to the party with the potential of secrets to unlock, he tugged a book from the shelf, as it appeared newer than all the other books, creases not yet broken into the bound spine. Fanning through the pages, he noted discussions and experiments for the creation of the stelgladios. He sank into the leather chair, plopping his feet onto the desk, and analyzed the information before him.

The book spoke of the metal being forged in such a way, with perfect temperatures at specific times to alter the force, giving it a magnetic quality. The Satelles used stelgladios to repulse shineblasts in an opposite direction. The charts and diagrams of the necessary elements caught Maddox's attention as he immersed himself in the pages.

He pulled an orange from his pocket he'd swiped from the table downstairs. He peeled it methodically, in one single piece, tugging the soft middle out before discarding the outer layer into the crumbled papers filling the can next to the desk. Of course, the waste would be indication of someone having been in the room, but he anticipated the staff who cleaned wouldn't necessarily think twice. He twirled the juicy sphere in his hand as he positioned himself to continue reading.

Flipping forward, he saw the constructs of the Satelle capes. Some new thread had either been found or created, and when woven into the threads of the cape, it allowed the final product to absorb shineblasts until the power of the charge faded away. This was progress for Lume. They had created something entirely new and different than anything he had seen either at Seclus or at home.

The door slowly groaned open, and Maddox shut the book to face the intruder, witnessing soft caramel waves fall into the room as Theodora peeked in. It had been days since the last time he'd seen her, not that it should have mattered. She closed the door softly behind her and instantly went from inquisitive to furious. Rage filled her green eyes, and he ducked his head down to contain his smirk.

"What can I help you with, Theodora?" He relaxed back, pulling slices from the orange, enjoying a tangy, sweet burst with each bite.

"What can you help me with?! What in the fates are you doing here? The last time I saw you, you were bleeding in the alleyway before being whisked away by three people I didn't know. And I had no way to find out if you were okay, or even alive."

"As you can see, I'm fine." He gestured to himself, sweeping away the emotion she poured into the room. He might have dropped his guard the last time she saw him, but he wasn't going to let it happen again. "Did you see the prince's shoes?"

She ignored his question completely. "You must have hit your head on the way to the ground because you forgot that only one of us was supposed to be here."

He shrugged. "You went after four men alone. I didn't know if you would be in any kind of position to attend." It wasn't anywhere near the truth, but it fit the situation.

"You know where I live, Maddox. You checked on me every day since Jove. You could have checked on me before tonight." Maddox continued tugging orange slices, refusing to fill the silence. "So, you show up, see I'm okay, and hide upstairs?"

Accusations entwined every word from her mouth. He *really* didn't want to play this game with her. He wanted to take care of what he needed and get back home. But every time he looked at her, the idea of what home meant slowly started to change. At first, it was those green fields he remembered as a child, the way the morning light would make the dew drops on the grass blades come alive. But now? It was the shades of green that stared him down, refusing to accept his lack of emotions, attempting to dig them out of him with a blade.

He got to his feet and moved to the front of the desk, slowly closing the space between them. "What do you want me to say, Theodora?"

"Let's start with the truth."

"What did you expect me to do? The last time I saw you, it was rather chaotic. I learned that you went after four men and wreaked havoc over Seclus. The Satelles have been on my ass trying to find out what happened, where you were, and my involvement. I showed up to see if you were okay. Once confirmed, I continued with the plan we agreed to: finding out if there were any secrets contained in the house."

"Why can't you tell me what you're thinking? Why can't you let me in?"

"I told you!"

"Cryptic and terse responses aren't telling me anything."

He wasn't doing this. He'd already let out too much. He began turning back to the chair, maintaining his composure, pulling himself back to things that made sense in the world. Cause and effect. Action and reaction.

"What are we doing, Maddox? We've always kept our work separate from each other. And now with Jove and my Seclus conflict..." But the fight in her words was already fading away. "What is it you want, Maddox?"

There it was. The actual reason for her trying to make an argument over something that wasn't even arguable. "You want the truth? I will give it to you one time, Theodora. One damn time." His voice didn't rise, didn't turn to yelling. Nonetheless, it increased with intensity. "I have always put myself before everyone. Me. My needs, my wants. My actions do not serve anyone else's purpose but my own. But you?" He cut off in a growl, looking everywhere else in the room but at her. He couldn't do it. He didn't know how to put these—these *feelings* into words.

When he met her face again, she did not balk. And although she stood shorter than him, it didn't preclude her from staring back at him defiantly. He leaned forward slowly, resting his forehead against hers. He closed his eyes and breathed deeply. The sweet citrus on his breath mingled with her rich woody scent. The warmth of her skin flowed through him, igniting something deeper than he ever expected. He hoped the words he thought in his head somehow made it to her.

"You do this to me," he whispered, and it was true. She frustrated him to no end. She did things in a way he didn't like; made decisions he didn't agree with. But he would rage against the world to make sure she was safe and happy.

The strain between them began to melt away. She pulled away from him, and he immediately wanted to close the space again, but she tilted her head upward, a smile breaking across her face. He hadn't seen her smile at him before, not unless he made a joke, and it felt like Fiedel finally rising in the morning, the light cascading over the horizon.

He leaned closer, slowly allowing their lips to graze each other for a small moment. It was swift and relayed none of how he felt about her. He battled with himself on whether to kiss her again, to let his walls collapse, when she grabbed both sides of his head and pulled them together. It was lips and tongue as the heat spread down to his groin. It was a dam breaking open, the waters flooding over him, destroying anything and everything in its path. All he could do was *feel*.

They separated from each other, both attempting to catch their breath. She stared at him, and he knew expectations had been what pulled them apart. He had secrets to find, and she had stories to weave among the Lumens.

He leaned forward. "I will always find you, Theodora." He kissed her cheek, sealing the promise.

Astrum

*C*hildren's giddy laughter and playful screams pulled Theodora from her sleep. After days of becoming complacent in her training routine, it was an inner battle between getting out of bed or curling back under the woolen blanket. The first cool astrum breeze snuck through the open window, making her strongly want to choose the latter.

She wiped the sleep from her eyes and tumbled from bed, silently chiding herself for staying holed up in her home. And for what reason? Maddox had been fine. Worse was the weight of everything else, the reality of what she had agreed to—courting a prince, killing a king—burying itself further into her bones. The grains passing through the sandglass became more audible as the days passed.

After stretching her muscles, she started bouncing on her toes, throwing out some punches. Two, three, three two, duck. Kick, kick. Two, three, three two, sidestep. Sweat formed across her skin, warming her blood as she continued through the sequences. Five, six, three, two. Her heart pounded in her chest as she shuffled over the wooden floors. Keeping her steps light and quiet, she felt the tightness in her core.

But no matter how much training she did, nothing could prepare her for the task at hand. Even consciously thinking of it made nervous bugs crawl around in her stomach. Wiping the sweat from her hairline, she turned to face the window, where the morning light filled the room in an orange glow, the fog since lifted with the new day.

There were always festivals, always reasons to acknowledge Lume's great king. One was guaranteed with every change of the season, each varying in their expectations. When the first snow of heim fell, the Satelles would build a massive bonfire in the middle of the capital in celebration. It would blaze all day and into the evening, and as families left to retire, they would take a burning ember to their homes, lighting fires of their own, a small spark connecting Lumens together. But it was by far the worst of the festivals for Theodora. It was too strong of a reminder as she watched families walk away together, the children begging for the chance to carry the ember home. A reminder of how much she had lost. Of how much she would never get back.

Fevron was filled with a cascade of color. The children were required to wear white on the day of the festival, a reminder of where they had come from and their innocence to the cruelty of the world. The adults wore black to mourn their pasts and the ancestors who came before them. But weaving through the sea of monochromaticity, the festival was decorated with bursts of bright flowers from freshly bloomed peonies, irises, primrose, violas, and hyacinths.

Lumens gathered around the castle and swam the depths of the Syrenic Lake for the Solta Festival. Anything to stave off the wretched humidity. Numerous wooden rafts of varying sizes would be placed along the blue-green water, with races back and forth from the lake edges and noodling for catfish along the shores. The competitions lasted throughout the fiedelight, daring Petram to take their world back.

She turned away from the window to wash and change before exiting her building. The season change was imminent in the air. The festival celebrating astrum had become the largest of the four. The changing of leaves, which Theodora anticipated would happen any day now, signified forgiveness of past transgressions. The acceptance of change. The entire capital was decorated in bright colors, and vendor carts displayed new foods for people to taste. The festivities now spanned two nights; the second night being added shortly after her parents passed. The first was filled with music and dancing, and on the second night, Lumens curled up with their loved ones to watch exploding flares in the sky.

With the warm air vanishing, Theodora could almost taste the excitement as Lumens prepared for the Astrum Festival. They bustled in every direction, some to vendors and shops for last-minute items, others to taverns in anticipation of the midday meal. Shop owners decorated lampposts with twinkling lights and erected banners along the walkways in hopes to garner more patrons. Theodora watched as children scattered picotee petals along the cobblestones. The capital was filled with a momentous amount of color, reflecting off the jewels hidden inside the buildings, tinting their world in warm tones.

Theodora approached Earleen's shop, nestled between the bookseller and horologist, the door hidden a couple feet back from its adjacent shops. Today a stand had been erected out front, showing off a glimpse of stunning fabrics. She brushed her hand along them, skimming across the different textures, sometimes catching on sequins and beading.

Her boots on the wooden floor announced her presence as she walked into a whirlwind of color, like a painter's tray. The inside of Earleen's shop was plastered with airy sky blue and accented with bold yellow designs painted onto the walls. Cords spanned from wall to wall along the ceiling, allowing fabrics to drape into the space. The fabric caressed her face and clung to her body as she searched for the tailor within, the sounds of the market square barely audible through the open windows over the thrumming gears of the sewing machine.

Earleen was hard at work over the machine. His gray hair fell over his face, nothing but his long, beaklike nose peeking through the strands. As he pressed his foot up and down, a series of gears caused needles to work in tandem, pushing thread together and through the fabric. He glanced up, pins tucked between his lips, and his hazel eyes peered at her through his glasses. The lenses magnified them from her perspective, making him appear more owl than man.

Recognition caught and he jumped up from his seat, smacking his head on one of the wooden shelves along the back walls, causing the pins to scatter from his mouth. "Ho, Theodora! I was waiting for you to show. I've pulled some fabric to show you for your festival dress."

Although his gray hair attempted to show his age, his face and skin did not. Other than his nose, his face was smooth and had a youthfulness that never faded. Not a single wrinkle graced his face, and even through the fiedations, she had never witnessed him age.

The chattering of women sounded from the front of the shop, followed by an immediate ringing of the shop bell, interrupting them. Earleen moved away from his workspace and beckoned Theodora to follow him, the limp which was gossiped about during Rajveer's party now apparent in his gait. They made way to a bulk of fabric lounging on a chair within eyesight of the new patrons. There were three of them, with full petticoats, cute bonnets, and parasols.

Earleen motioned to the squabble, indicating that he would be with them momentarily before turning his attention excitedly to the fabric. "This is the fabric I picked out. Let me deal with them and then we can talk about it."

Theodora couldn't stop the small smile from appearing. His wonder of life was contagious. As he went to deal with the three women, Theodora turned the fabric in her hands. It was soft, a deep sapphire with a bolt of black lace, and the color changed slightly in the light. It was so rich that it engulfed her, making her want to swim into its depths.

Earleen knew of her deep appreciation for his work. She allowed him full reign of his creativity and for her to be but a canvas for his art. Instead of a list of requirements of what she wanted or expected, she had always allowed him to create as he wished. The only request she ever made was to allow practicality.

She turned back to see Earleen pass the women their requested dresses. Always the same: huge skirts, ruffles, and shades of pink. Nothing like being sheep within the world of men. Although Lume and Seclus were similar, Lume was much more founded in tradition, all fabrics of full skirts and tight suits. Not a weapon gleamed on any person, although she knew a handful carried concealed ones. On the contrary, Seclus was much more practical, for Seclusians paraded trousers and boots for their work and their long treks through dirt and mud.

Earleen limped back to Theodora. Her eyes caught on his gait once more. "Are you okay?"

"Oh, absolutely! You mean this leg? Not a worry. Just some soreness from sitting at that machine all day."

He started to reach for the bolts to explain his grand idea, when Theodora interrupted him. "Earleen, we both know that's not true. At the prince's party last night, people talked about it. We're worried about you. Is there something I can do?"

"Really, Theodora, it's nothing. Just a small misunderstanding." He pulled at the fabric once more, motioning for her to strip and to approach the fitting stand.

She grunted in acknowledgement, stripping off her accessories down to her intimates. "Unfortunately, Earleen, this festival isn't going to be my usual. I am attending with the prince and need something appropriate."

"Theodora, do you think so little of the gossip of Lume? I had already heard of this. Please, step up."

She padded her sock-covered feet onto the stand. Earleen was quiet and quick with his work. His hands flew about her body with the measuring tape, tousling fabric back and forth before pinning it into place. Earleen had outdone himself with this outfit. The blue dress was bordered by black lace at the feet as well as along the plunging neckline. The sleeves were long with layers flowing from the elbows, capped with more black lace.

"What do you think, missus?"

"It's absolutely stunning, Earleen."

"I'm glad you enjoy it. I couldn't make some unique festival trousers for storing your gear, but I thought in this dress, I could add pockets to the sides?"

"That would be wonderful."

"Can I see your dagger for measurements? I'll have to make the pocket deep and narrow so that it tucks into the folds of the fabrics and remains discreet."

She nodded in the direction of the blade stacked on top of her discarded clothing. After taking the measurements, he looked back up to her. "Alright, well, out you get. I'll finish this up, and you can stop back this evening."

As she tugged her clothes back on, he situated the fabric onto one of the wooden mannequins.

"Thank you again, Earleen!"

"Certainly. I can't have you going off and finding a different tailor," he finished with a wink.

"You haven't let me down yet," Theodora said as she pulled klaud from her pouch and offered it to Earleen.

"I can't accept that," he said as he began turning away.

"For your trouble, Earleen."

"It was no trouble at all, but I appreciate the gesture."

"Thanks again." She nodded to him, slipping the klaud into his apron as he turned back to his work. "I'll be back this evening."

Already lost back in the fabric, he tossed a wave over his shoulder.

∴ ∵ ∴

Theodora headed back out into the market, the scent of bread catching her attention. She turned to see the bakery staff setting up stands outside, lining them with rolls and croissants. Twisting around Lumens in the market square, a gruff voice called out her name. She turned in the direction of the sound and noticed Rabb standing in a doorway, a thick burly hand waving in her direction. Adjusting her course, she greeted the butcher.

"I saw you walking and needed to ask a favor." Rabb was a tall, muscled man that more closely resembled a bear than a human. Dark hair coated his arms, which were visible, as his sleeves were shoved past his elbows. A white apron fell over his clothing and was decorated with blood.

"I'm headed that way. What do you need?"

"Can you take some pork belly over to Albani? He's expecting it, but I can't get down there with all this stuff for the festival."

"Not to worry, Rabb. I'll put it on your tab."

He huffed out a laugh and a thanks before he turned to instruct his apprentice on the payment: some meat in a couple days after the festivities had died down. With the wrapped pork belly in hand, Theodora detoured for Digere.

With her recent Seclus tasks, it was a reprieve to help Lumens. Barters of favors for goods. A much simpler exchange. As a child, she had merely wanted to help. She wanted to see, and be a part of, a majestic capital. And maybe because her parents had been local vendors, she felt she was continuing that legacy.

Her parents had owned a vendor cart, which her father had dutifully named after her mother, Imaginings by Isadora. It was such a bizarre name among the simple shop names of Rabb's Butcher Shop, Earleen's Fabric, and Amabel's Sweets, but her mother had loved the name, and they'd worked hard to one day own a store front.

Tears sprung to her eyes, and Theodora shook away the memories. She had been so content to continue with life through exchanges, but Seclus grew too much too quickly, and the opportunities were bountiful. She was stuck choosing between survival and loyalty. When Theodora first traveled down to Seclus, she'd been amazed at all the things her parents had been helping with and had loved watching them come to fruition.

Theodora dropped her face as she turned onto the alley behind Albani's tavern. She took a moment to take in deep breaths, willing the tears to dissipate, forcing the vivid memories away. She didn't find the right door immediately, because the one she needed had been left ajar. She knocked hard on the door frame before stepping in to locate one of the workers.

Theodora wanted to call out, but with the door left ajar and no workers about, something felt odd. She placed the meat on the nearby table and crept deeper into the tavern in search of a worker. She heard nothing, an eerie calm pressed heavy on her shoulders. Unsheathing her dagger, Theodora continued forward, keeping her steps slow and quiet on the floorboards. The empty tavern was further down the hall, the lighting dimmed low. The wood beneath her feet made a low whine as she placed her ear against a closed door and heard a long moan.

She stepped away from the door and brought her foot up, kicking it near the handle. The door burst open, swinging along its hinges, to slam against the adjacent wall. Her dagger unbaled to protect her against the embarrassed flush that ran up her skin.

Albani jerked himself up right from the chair from which he had been sleeping. The little hair on his head was pushed in dueling directions as he wiped a hand down his face.

Theodora reined in her surprise, but still felt the need to explain herself. "I'm sorry for the interruption. Rabb asked me to deliver some meat, which I left on the table." Albani continued staring in disbelief, his eyes decorated with sleep. Theodora continued, as if she might reconcile the situation. "I heard noises and thought someone was hurt."

Albani's loss of words hung in the air. Tense.

"Erm, so, I'm going to go, but please, continue." It was the fastest she had ever left Digere, and that was something.

As she walked the capital streets, she immediately chastised herself. Fates above! How could she had been so absentminded? As she left the cobblestones of the capital and entered the woods, she gave herself one final reprimand and then she moved on. As soon as her foot touched fresh soil, she broke into a run, hoping the distraction of dodging downed trees would occupy her mind.

Fighting against the pull of the branches, she broke out of the trees, the fiedelight blinding her briefly, and slowed down in one of the old farm fields. The previous farmer had long since left the world. Now the field was abandoned by his king and people. The dry, golden grass swayed in the tickling breeze. Off in the distance, a farmhouse barely remained standing, its windows blown out. It looked like it might fall if the winds were strong enough.

She let out a breath when she heard a twig snap from the woods behind her. Theodora swung her body into a crouch, pulling her dumgun from her boot and surveying the tree line, Albani's embarrassment now long forgotten. Four shadows stepped out of the trees and made no attempt at hiding their noise as they left the cover of the forest.

worthy

Sweat dripped down Rajveer's back, drawing attention away from the stelgladio in his hand. He brought a hand upward, forcing the training to stop for a moment as he tugged his shirt off. Wiping his face with the fabric, he tossed it in a heap into the corner.

"You know, this is why your training sucks, Raj."

"Shut up!" Rajveer got back into position, the stelgladio held tightly within his hand. When he finally gathered himself enough to start the training again, Amicus continued.

"Do you think an assailant is going to stop for you to take your shirt off?"

"I was sweating."

Amicus glared at him. Rajveer had kept his trousers on, but his boots and socks were long since discarded. Meanwhile, Amicus continued the entire exercise in complete Satelle garb, cape and all, his locks loose about his face. Rajveer thought Amicus' loose hair would serve to his advantage, another irritant for Amicus, yet he hadn't even attempted to brush it aside. Fates above, Rajveer thought he would have at least tied it back by now.

Fiedelight entered through the tall windows and balcony doors, reflecting off the lampposts standing watch along the white walls illuminated in gold. Staged in one corner of the room was a rack of various blades, while the other side of the room held a fireplace. He still didn't know the reasoning behind his great-great, or however many *greats*, grandfather's request to have one in the training room. Even in the coldness of heim, when most of the other rooms were threatened by the chill, they never saw any need to use it.

Rajveer looked at his friend, his training experience clear in his relaxed stance. Amicus held his blade loosely, gauging and calculating every move Rajveer made, waiting for the small cues in the twitch of his muscles.

"You've been doing this a long time."

"I won't be offended at the insult to my age. But regardless, you are going to be king in a couple of days. None of this is going to be easy, Raj. You need to accept that."

"Talk about shedding light on the impossible task still ahead." Rajveer hesitated a moment before continuing, "But I don't know if we have to kill him."

Amicus stopped his swing mid-flow and lowered the blade, his brows pushed together. "What do you mean?"

Rajveer took advantage of the dropped defenses and exerted a variety of attacks, but Amicus barely put forth any effort to stop him. Rajveer choked down breaths as he explained what Miles had suggested.

"If I marry Theodora instead,"-he shrugged, letting the thought steep for a moment- "then we might not need to kill him. He only wants to see me marry. It's the only thing preventing him from stepping down."

"So, you think if you marry a Lawless, your father will give up his crown? Fates above! Do you guys ever think anything through?"

Amicus sheathed his stelgladio, rubbing his hands up and down his face, as if he could wipe away Ravjeer's questioning.

"Why not? He has said numerous times that the only thing stopping me from getting the crown is that I haven't wed. He was ready to hand it to me with Alouette…"

"Raj! Why do you keep doing this? Why do you keep coming up with reasons why you shouldn't kill your father? Did you seriously forget what happened to your hand? Because every time *I* see it, it is a constant reminder. Your father isn't going to let you marry Theodora."

"He doesn't have to let me do anything. Why are you so fixed on killing him? Why is that the only solution for you?"

"He isn't going to abdicate the throne!"

And there was the crux of it all. Because although to the rest of the world his father was complacent, his father had always kept his nose buried deep in Rajveer's life. Rajveer didn't know what his father wanted. He looked down at his left hand, as if it might provide some insight. Some damn clue on what to do.

He had been born with a left hand, but the memory of what it was like was long since burned away by the pain of having it removed. The pain bore into him now, the constant dull ache from the metallic claw, wrapped around and attached to his wrist. It remained every day. The long external rods that worked up his forearm before diving back into his body were connected to his bones and tissues.

He hated reminders of the past, and here a permanent one resided for all to see. He looked back to Amicus' stone face, regret etched across it. They hadn't spoken of the incident except for immediately afterward, when Amicus came to him begging his forgiveness.

"You know that was the turning point, Rajveer. The king even asking me to do that was outlandish. And then doing it? Every day I watch his back, guard his livelihood, but do not think for a moment that I wouldn't step aside if it meant protecting you. Right now, you both hold the same end, and it's easy to keep the mask in place."

Amicus set free a sigh, and Rajveer felt it burn deep within his chest. He looked back down at his hand.

"I don't envy you, Rajveer. If I could take this weight off your shoulders, I would. There is no reason to keep going back and forth on this. You need to figure out what it is *you* want. And no matter what that is, I will stand guard by your side to help you get there."

Rajveer thought for a long moment. What *he* wanted? He had never been given the opportunity to dwell on it. Not to say that he didn't think of it, especially during boorishly long dinner conversations, but he always saw it as out of his control.

He filled his body with an air of confidence and tightened his hold on the blade. "I want to take the throne and become something more, to change Lume and become worthy of all their stories."

"Then let's get to work."

Lawless

M addox walked out of the trees, both hands deep within his pockets, Valix and the Gems surrounding him.

"You know I could have killed you all," Theodora shouted across the field.

She was adorable when she acted so flippantly. He chuckled quietly to himself. "Don't flatter yourself."

"Now you're following me?"

Her weariness was apparent in her eyes. Although her body and face aired boredom, she didn't realize how much her eyes gave her away.

"You told me to stop brooding in the dark." A shrug pulled at his shoulders.

"I never said brooding," she called. "And why are they here?" She waved the dumgun at his companions as they all approached Theodora, closing the distance.

"Since I know you can't stand secrets, I figured I would introduce them to you." He paused, deliberately looking at the metal in her hands. "Put it away, Theodora. No one is going to hurt you."

She tossed him an irritated look, rolling her eyes before securing the dumgun back in her boot. Maddox slowed, and they stopped in a loose circle, a group of Lawless to any outsider. Valix and the Gems were dressed similarly to Theodora, trousers and tunics strapped on with leather and gear. Maddox remained out of place, like royalty among them, with his crisp buttoned shirt and vest. No cravat today.

Maddox pointed to each of them for their introductions. "Valix, Gemma, and Jemmie."

She nodded in acknowledgement at each of them before confronting the twins. "There is no way I'm going to tell the difference between you two."

They turned as mirrors, showing a darkened mark on their skin below their opposite eyes.

"A convenient birthmark, it appears." That was seemingly the only difference between them, as their personalities remained the same. Stoic, unemotional, and robotic.

Theodora continued, turning to Maddox. "Do they not speak?"

The twins faced Theodora and spoke in unison, "There is confidence in silence."

Maddox let their words hang in the world, the breeze coming to taste them, before he spoke. "They do not see the need in speaking often. To them words can be used for comradery and love, but they can just as easily be used for betrayal and deceit. Without the misdirection of words, actions speak undiluted truth."

"And what about him?" She pointed to Valix. "I know he speaks."

"He does," Maddox replied for him, "but he is my second. He is much more inclined to observe and only speak, when necessary, especially when I am present."

"Impressive...and you are an entourage?"

"We are considered a committee, or more specifically, the Second Committee."

Maddox saw his companions shift their gazes discretely in his direction. He could feel their questions burn into him as he gave this information to Theodora.

"Are you saying you're not from Seclus? I've never heard anything of committees before."

"We are. We are one of the ruling committees of Seclus."

"So, who is the first?"

Silence swarmed about them. Even the birds and the insects steadied their movements, listening intently to the answer.

"What are you doing out here?" Maddox questioned as he gazed across the field as if the answer would present itself.

Theodora scowled at him, probably from his lack of response. "I'm here to see my horse."

Maddox gestured to the others, and his three brethren turned away, making themselves comfortable within the shade of the trees. He motioned for Theodora to continue with getting her horse. She stepped away from their group and whistled. It was an odd tune full of rhythmic jumps, and nothing close to ordinary. The sound blew across the field, and a breeze responded by caressing them.

After a few moments, the sound of hooves vibrated through the soil and the stunning mare Maddox had witnessed a handful of times came into view. He watched as the mare neared, bumping hard against Theodora's body, and he witnessed her relax with the interaction.

"What's her name?"

Theodora's face scrunched up. "Um," she hesitated. "I call her Down River."

"Down River? That's an odd name for a horse."

She shrugged. "When I found her, I had barely crossed into adulthood. I was walking through one of the fields and started whistling the song. *Down River?*" Maddox shook his head but let her continue. "She just showed up. And when I sang it, she followed me. Since I don't own her, I didn't want to give her a name of my own. But she comes when the tune is sung, so I figured I needed something to call her in my head."

"Are you going to sing it for me?" The smirk instinctively formed, and Theodora scoffed and stared back at him. The horse named after a song jerked her head back and brought it down quickly with a snort, mirroring Theodora. Maddox involuntarily flinched and turned his body slightly away.

"Are you afraid of horses, Maddox?" Theodora grinned.

"I wouldn't say afraid. I prefer not being around them."

"Why is that?" Theodora continued stroking the mare. Her snout was buried in Theodora's hair.

"They're unpredictable. I prefer knowing what is to be expected: a cause and effect. Horses are the antithesis of that. It is in their nature to be fickle creatures."

"It suits you," she responded as she pulled away from the horse to face him. He pushed his eyebrows together, allowing for the question to show on his face. "Being predictable and logical, I mean. That makes sense. It's all very much you." She paused a moment before continuing, "Do you want to pet her?"

An internal debate teetered within him. She extended a hand and inclined her head, beckoning him to take it. And Theodora had been right. He preferred life as an analytical equation, one where there were set variables and one could use reason to search for the unknown. But she made him *feel*—hopes, desires, far-fetched whimsies, and the illogical. With taking this small leap with her, he could face some small dread. The song-named horse was wild, even more wild than the ones the Satelles used, and yet despite his fear, he also wanted to feel Theodora's hand, to touch her skin, to caress its smoothness against his rough hands.

So he took it, their calluses grazing against each other. The warmth of her flowed up his arm and fell deep into his core. The lyrical horse continued her picnic through the dried grass while Theodora lifted their hands, together, gliding them along the side of Down River's muscled shoulders. Maddox felt the pure strength of the beast beneath his hand, the rush of power surging through him.

Time seemed to stop for a moment, the worlds aligned for a brief pause. There was no awkwardness or tension in the silence that fell, just peace. Power and undiluted peace. Maddox pulled his hand away and tucked it back into his pocket.

"Why do you use a dumgun?"

The curiosity in the question scrunched across her features, most likely due to his randomness. However, it was a question that had bothered him for some time. Almost all Lawless had moved to shineguns, especially those in Seclus, where dumguns were now considered old and obsolete since a dumgun had to be reloaded after four shots. The shinegun could rally a countless number before it lost its charge completely. Even the Satelles didn't use dumguns anymore. And yet, it was Theodora's preferred weapon.

"If I am going to hurt someone, I want to mean it and make it count. Just like you believe Down River to be unpredictable, that is what a shinegun is to me."

Well, she wasn't wrong. Shineguns tended to be finicky. There had been several times when a Lawless didn't properly charge the blast and people were either burned, or it had ricocheted off bodies completely.

"Here, let me see your shinegun. I can alter it for you, which might make you feel more comfortable with it."

"Alter it how?"

"Well, the problem is that the shinegun loses its charges after a time, even when it isn't being used. I found a way to provide continuous power so that the charge doesn't falter. You can shoot blasts, make explosives…" he said with a smirk, and he held out his hand, awaiting the piece.

She knelt forward, her hair falling about her face as she pulled the shinegun from her left boot. The discs were already starting to dull from the loss of their charge. He immediately went to work, pulling a long cylinder from within his jacket as he took the bottom off the grip of the shinegun. Wiring fell out in its place, and he gestured to Theodora for her to hold out her hand as he placed the end into her palm, focusing on the disrepair in front of him.

"Instead of the source of the power running from the charging discs, I'm attaching it to this alternative source. I've had a couple made. Some didn't do well before, but these seem to be holding up. They don't last forever either, but they allow the shinegun to go without the fiedelight for a good period and not lose any charge."

He slipped into his methodical pace. The one he found the most comforting. The world was chaotic. It was full of emotions, the politics of power and who held it, and the business of earning klaud, legitimately or not. But tinkering with an object? It was merely the need to solve a simple problem, one that had a definite solution.

He finished up his work, taking the grip from Theodora's palm before sliding it back into place. From the outside, there was no difference except for the metal discs. Even now, they had turned darker and shinier in color to indicate its full charge. He handed it back over as her lips lifted into a small smile.

Maddox turned to Valix and the Gems, where they were stationed in the shade. He appreciated these small moments with Theodora, but he knew they were watching him intently, probably analyzing his actions. Not that he expected them to do anything differently; it was what they had been raised to do. He would need to leave soon. There were other items he still needed to address today.

"Are you going to the Astrum Festival?" She interrupted his thoughts.

"I am. I would ask if you were going, but…"

She laughed. He took in a deep breath, letting the joy fill his chest. He wanted to memorize that sound. A smile formed on his face against his wishes, his heart splintering at the way she looked at him. "Have you eaten?"

She shook her head. "It's a busy life for a woman preparing for a festival, especially one with the prince. You know, full gowns, dozens of bonnets, parasols galore." Each word grew grander, her arms splaying wide across the field.

"You don't have a bonnet or a parasol."

"Oh. And you are certain I have a full gown?"

"If you weren't going with the prince, I wouldn't be so certain. But since you will be decorating his arm, a gown is without question." She looked down to her feet. He didn't want to keep talking about the prince or her going with him to the festival. "Let me take you for food. That will be enough for me to celebrate astrum."

"Just as Lawless?"

"Just as Lawless."

Although he repeated it to her, it broke apart a small piece inside him. He extended his arm out for her, a slight bow before she intertwined hers.

They walked to where Valix and the Gems remained posted. All three turned in their direction, Theodora slipping her arm out from his as they got closer.

"I'll meet you at Rosa, charm."

Theodora looked at him quizzically for a moment but thankfully said nothing, instead beginning her navigation through the few trees back to the capital.

"What are you doing?" Valix asked when Theodora was out of hearing distance.

Maddox avoided providing them the finer details. "I'll have a portal open here in a few minutes for you three to return to Seclus."

"You're going with her? To do what?"

"Valix, I gave you your instructions."

"Fuck that. I'm your second..."

"And as my second,"-Maddox turned his body, facing Valix fully, looking down with the few inches in height he had on him- "you are to do as you are told without question."

"My job, sir, is to question when you're making irrational decisions."

"None of my decisions are irrational, Valix. Is that clear?"

"Understood, sir." Valix lowered his eyes, obedience granted.

Maddox shifted his gaze to the twins, raising an eyebrow in their direction to question whether they were going to defy his instruction as well.

"Yes, committer." They said in unison.

Committer. His title was an outward reminder of everything he was focused on. All his obligations and decisions came down to it, and right now, they all rested on the Lumen king's assassination.

Nightmares

Theodora walked with Maddox arm-in-arm through the capital, feeling as if she had caught butterflies in her stomach, the fluttering of wings deep within her core. She tried to tell herself that he was merely acting the gentleman enjoying the release of solta, but the back of her mind overanalyzed everything. The gray clouds swirled above while Fiedel, with its orange crown, slipped to the horizon. As they walked silently, she reflected about their dinner: how he had swirled his beer, how he'd said her name across the table. Did the kiss they shared nights ago mean anything?

They had detoured on their walk back to her home to pass the small number of vendors who remained before heading to pick up her dress from Earleen. Most of the vendors were beginning to pack their carts, collapsing boxes, and breaking down the awnings. The light song from a fiddler harmonized with the cascading water in the fountain.

A man lounged against his cart, making no move to start closing it. From it hung ropes of bulbs on dowels in various colors of pinks, reds, and yellow. As Theodora continued her path forward, she noticed the man perk up, rising quickly from the ground.

"Good evening, my Lumens." He bowed low, and his pointed, deep purple hat threatened to fall off. Theodora caught the twitch of Maddox's muscle along his temple in her peripheral at the name bestowed onto him: *Lumen.* "Can I interest you in a jelly pop?"

Theodora had seen the cart here before, Clo's Concoctions, and had been curious about the boisterous man with black hair cut straight below his chin. During the day, he was popular enough, as long lines had prevented Theodora from venturing over to witness what exactly he sold. Theodora inclined her face to Maddox, raising an eyebrow in question. He merely nodded back, letting her make the decision.

"What are they?" Theodora asked as she carefully reached her hand out to a pale pink one. She wondered what was inside and lightly slid a finger along the translucent shell— but shell was the wrong word because it wasn't hard. It was a soft film, almost gelatinous.

"They are a sweet delicacy! It is fruit juice wrapped in translucent seaweed. When you bite one, you are left with a surprise."

She hesitated a moment as she continued to stare inquisitively at where they hung. "We'll take two."

The man pulled two pale pinkish bulbs from their rope. Maddox fished klaud from his pocket and slid them on the tray.

"Thank you for your business," the man responded as they turned away from the cart and continued to walk along the cobblestones.

"Two?" They had barely made it out of ear's reach of the vendor before Maddox was questioning her.

"Fates, save me," she whispered to the sky before turning back to Maddox. "You think I am going to eat this by myself? Here." She proffered one and kept the other close. "On the count of three. One, two…" She trailed off.

A soft chuckle slipped past Maddox's lips before he tossed the jelly pop into his mouth on her count of three. She mimicked his movement and bit down on the soft candy, where a juicy burst exploded within her mouth. The taste of strawberry was immediate.

A moan caught in her throat as she swallowed the sweet, and Maddox nodded in agreement. When their eyes met, he smiled at her, and she had to choke down a laugh. His teeth were a glow of pale pink, and when she opened her mouth to tell him as much, his eyes widened at her. Clearly the surprise wasn't from the juice but from its aftermath.

"It must be whatever is inside that makes our mouths glow like this."

The two of them immediately went back to the seller, Clo. Although he was happy to be earning more klaud, a puzzled look laid on his face. It made them laugh even harder, the two esteemed Lawless brought to tears over something so childish.

Theodora wiped tears from the creases of her eyes, waiting for their laughter to finally calm and the glow of their teeth to fade before heading for Earleen's shop. They continued lazy circles around the capital, and Theodora wished the night would never end.

∴ ⌄ ∴

The blue fabric of the dress Earleen had created hung from Theodora's closet door. Her mind kept reverting to the memories of last night with Maddox, albeit a mostly uneventful one according to Lawless standards, there had been a thrum underneath it all of something more.

He had walked her all the way up to her door like it was some typical courtship and not a night spent together as *just Lawless* like they had said. He even promised to return this morning to help her prepare for the festival. She couldn't believe it was here already, the days having dwindled faster than she could imagine, but now time seemed unhurried.

She forced down some eggs and potatoes, but they did nothing to help the nerves that flitted in her gut, like a flash crystal about to explode. After a bath and an immense amount of brushing knots out of her hair, she left her hair loose, cascading about her shoulders, and swept rouge across her eyes like her mother had shown her. None of the tasks were distraction enough, so she stood in her soft robe staring at her gown, finally lost in the waves of fabric.

There were few times when Theodora would succumb to her memories, but this moment had stalked her since she had seen the bolt of fabric in Earleen's shop. It was the same color, a blue as deep as the ocean depths, of her mother's eyes before she'd left this world. Now, as she stared at the completed design, flashes of that night, a night so long ago she had hoped the fates or time itself would erase it from her mind, played out in her head.

The sound of the door as the wood broke away from its hinges reverberated through her bones, along with the noise of footsteps pounding into the floorboards of her once-home. The men, if that's what they were, wore black from head to toe, including black metallic masks of piping and gears. The masks had haunted her for countless petriks, but she had never seen them again.

No matter how she searched, as both a child and a Lawless, she was never presented with a clue about who they'd been. Despite her attempts to find answers, or hints, or *anything* as she replayed the scenes repetitively in her head, all efforts were futile. Any indication of the men's origin had been lost to the world.

Yet, the part that stayed with her the longest, the one memory that failed to fade, was the echo, echo of the two dumgun shots, one for each of her parents. She saw the bullets as they entered their skulls, so quick, their heads and necks snapped back. Her mother could barely mourn her father before she received his same fate. Blood flowed from the wounds, down into their eyes, following the bridges of their noses to their mouths, before they collapsed to the floor.

The shock of what her past-self had witnessed left her unconscious, tucked away under her parents' bed, until her eyes pushed open that morning, her parents dulled, lifeless bodies in front of her a reminder of her new reality. That night when she had been knocked out was the last Theodora ever slept peacefully. Now her slumber was filled with the physical pain of gripping her dagger and the emotional pain of waking up countless times from either the sounds around her or the nightmares within.

A knock startled her from the past, and she quickly wiped the tears that teetered in her eyes. She pulled the robe more tightly closed as she cracked open the door. Maddox was standing in the darkened hallway, his hand tucked behind his back.

"Hi," she whispered, as if all the notions between them might flutter off if she spoke too loudly. She opened the door further, letting him enter the space.

He pulled his arm around, and his hand was filled with flowers. "These are for you. Happy astrum." The sound of his voice awoke the butterflies in her stomach.

"They're beautiful. Thank you."

She took them, smelling the array of burgundy and orange petals with the occasional pop of muted gold-yellow. Their wild scent still clung to them as she went in search of a vase. After an awkward, ineffective hunt, she settled on an old boot from the closet.

"I always forget how lovely astrum is until the leaves start changing. It seems every fiedation it gets more and more beautiful." She spoke over her shoulder to where Maddox had stationed himself in front of the window as she fiddled with the stems.

"Only you would find beauty in the world dying."

Theodora snorted, forcing Maddox to face her quickly, raising an eyebrow as if to ask if she really did just snort. She rolled her eyes and went back to the flowers, pretending to obsess over them. She watched him from the corner of her eye, in awe of how out of place he appeared in her home. He was unbefitting everywhere he went, like he didn't belong to anything in this world. Even now, as he peered down at the streets through the tattered curtains.

"Are you ready?"

"Obviously not." She indicated her robe.

"I thought I was wrong on the gown part. I mean, dressed like that is one way to sway Rajveer."

She let out a scoff as she took the dress from the door and moved to the washroom, shutting Maddox out. She dropped her robe in a pool on the wooden floor. She pulled the bright blue dress over her body corset, the sleeves flowing to her wrists as she moved to clasp the front hooks closed. She pulled her stockings up from her toes, reaching the tops of her knees before she strapped them into place with her corset. She refused to glance at her reflection before opening the door and crossing the threshold.

Maddox appeared to be carved from stone, no emotions etched in his face. His inability to be affected was unnerving, especially when her world was spiraling out of control, giving her no way to ground herself.

"I'm glad to see I left you speechless." She threw the words out like she was putting armor over her heart.

He made a gruff sound, and she wasn't sure of the motive behind it. "Where do you put your gear in all of that?" He had a look of incredulousness.

Theodora pulled a chair away from the table, placing her boot on the seat. "Usually, I don't have this problem because I only wear trousers, but…" She grunted as she forced her heel into the boot and began tightly pulling at its laces. "Earleen sewed pockets into the gown for my dagger and some of my gear. They aren't big so I have to be mindful of what I take. At least I have a thigh holster for my dumgun, so that helps."

Maddox picked up the holster she mentioned from the table as she finished with her other boot. "Do you need help?"

She hesitated a moment, her booted foot still on the chair. She pushed aside the hair that had fallen into her face. "Erm." She had to clear her throat and force her voice out. "Okay."

It was barely audible, and at first, Maddox didn't move, making her wonder if he had even heard her. She could put it on herself. It would be difficult with all the fabric she wore, but not impossible.

When she opened her mouth to speak again, Maddox knelt in front of her. She had to suppress a jump when his cold fingers met her thigh, the smooth leather brushing along her skin, igniting her core. She looked up at the ceiling, focusing on slowing her heart and relaxing her breathing.

Maddox clasped the metal together when the door flew open, and Danika barreled into the house. She gaped at the two of them, her eyes darting between Theodora and Maddox, as if the answer to her unspoken question was going to be written across their faces for her to read.

"What are you doing?" Danika's voice thick with aggression at Theodora.

"Excuse me? This is my home." But even as she said the words, her faced flushed, whether from Maddox or Danika, she didn't know which. Maddox rose calmly, appearing oblivious to the entire situation. He turned to reach for her dagger, as if he was going to finish this right now. Theodora swiped it from him and threw her leg from the chair. "What do you want, Danika?"

"He isn't a part of this." Danika jutted a finger in Maddox's direction, completely ignoring the question.

Theodora took a deep breath. She was about to speak when Maddox interrupted, taking her hand. "I'm going to leave. We can't have two Lawless arrive at the festival together. I'll be around though."

"Thanks," she responded.

Maddox's hand slipped from hers. As he left, he eclipsed Danika and peered down at her, unspoken threats and a promise passing from him to her before he crossed the threshold. Danika slammed the door behind him.

"What is Maddox even doing here?"

"It's none of your concern."

"What's that supposed to mean?"

"None of this concerns you. Drop it, Danika."

"Well, it kind of does. The Satelles are working with you. So, why are you working with another Lawless?"

"He was there at the initial meeting. How does it *not* involve him?"

"We never included Maddox in the festival plan. I mean he was there when it was discussed, but he has no place here, especially when this is a Lumen festival, not some Seclus merrymaking."

"You do know Seclus is part of Lume."

"Not yet!" Danika sighed, rubbing her forehead in apparent frustration. "Look, I still care about you, Theo. We know nothing of this guy's past and you're throwing all your trust in him."

"I know as much about him as I do you."

Danika let out a frustrated breath. "Just be careful."

"Why are you really here?"

"I came to tell you I'm on duty for the festival tonight. I didn't want you to feel concerned about being alone." The words were casual, and less fight filled her voice now.

"I was anything but concerned about it. I've handled myself in far worse situations; I can handle myself against a meager prince."

The silence wedged between them, driving them further apart. Danika lowered her head, her voice erasing the last of any malice she'd previously carried. "Do you need help with anything?"

"No," Theodora responded quickly. "Just go and find your prince."

Danika turned to leave, and Theodora slipped her dagger into its holster. She heard Danika shuffle to a stop at the doorway and crossed her arms across her chest, waiting for whatever plea Danika would make next.

Danika twisted back to her, "He is your prince too, Theo."

Ablaze

Rajveer and Amicus strode out of sitting room where the tailor had finished the hem of Rajveer's trousers. Rajveer tugged at his jacket collar as he attempted to adjust back to the extra weight from his formalwear cape.

"They act as if we continue to go through growth spurts."

Amicus chuckled before shaking his head. "Danika should be meeting us at the stables. Miles and Jude are preparing the horses now so we can head to the capital."

"Danika?" Rajveer couldn't hide the annoyance that fell from his mouth. "I said I didn't want her there with Theodora present."

"Raj, you're being childish."

"No, Amicus. I told you, she still has feelings for Theodora. You can see it in her eyes when she talks about her. I don't need that interfering with the ruse we are about to put on for our people."

"I talked to her; she'll be fine. You can't stop trusting us now. It's about to get chaotic and you must keep your confidence in us."

Rajveer rolled the words off his shoulders, stretching his neck as if the whole situation were a physical annoyance he could manipulate away. He had no other option. Okay, not true. He could have married some wench and not be in this predicament. His father valued such old Lume traditions, but he'd made marriage a political game instead of a sacred vow, a vow Rajveer only wanted to make to Alouette.

The truth was that this had to work. The Lumens knew Theodora already and spoke to her credibility. Yes, she bent the rules a little bit. Yes, she had threatened their lives before. Then again, she'd also helped the local butcher deliver meat, aided in the setup of a local vendor cart, purchased goods from the outskirts of the capital for the temple, and carried a child through the rain for a mother of five. The Satelles had admitted their hesitation regarding her was due to her openness about being a Lawless, her big boots and gear decorating every part of her, and not with her character.

After meeting her himself, Rajveer hoped that she would be the bridge to connect Seclus and Lume together, because although on the outside she appeared Lawless and she occasionally committed some crimes, he could tell her heart still belonged to the capital. Not to the king or the crown, but to the Lumen people.

Thinking of Theodora... "Shit. Do you think she is going to dress appropriately? We didn't give her any guidance for her gown?"

"Raj, she will be fine. I'm sure of it."

"Why didn't we have her use our tailors? What if this—" His worry disappeared as his Satelles caught his attention, their laughter emitting from one of their rec rooms, Rajveer to take a closer look.

The room was filled with various tables and settees containing a handful of his father's guard. Most had their jackets discarded on hooks along the wall. The scene struck a nerve, not because of the mere fact that they were lounging about when a festival was imminent—his father would be well-guarded, even if these handful of soldiers were slightly incapacitated—nor because of their laughter. Rather, it was the slurred words that fell among them.

He felt neglected. It may have been less than a day since the last time he'd drank, but it felt like a lifetime. The mere fact he had remained sober through the entire dressing of the festival was miracle enough.

Rajveer stepped into the archway, taking in each person, their flushed cheeks and heavy eyes, before his eyes landed on the decanter in the middle of the table. They were drinking something, and it didn't look to be Lume's wine.

"What is this?" he asked, entering the room. Amicus hovered nearby.

The few guards with their backs to him turned to look at him, all of them still with a faint smile playing at their lips. A stockier man with a trimmed beard, whose name escaped Rajveer now, lifted the decanter, showing the clear liquid inside, flakes of gold swirling about.

"Just the right person to show up. Come, prince, join us in trying this new drink."

New drink? He didn't know if he'd asked the question out loud, but the Satelle continued.

"Our Clara here,"-he laid his arm over another's shoulder, her grin as wide as it was mischievous- "was doing a quick sweep near Seclus today. A boy was leaving with this decanter here. She stopped and questioned him. He said he was taking it to his mum. When she asked him about it, he said it was some new drink he had found out about recently. He called it...what did he call it?"

"Vocatus," Clara interjected. "A new drink the Seclusians had and were keeping for themselves. We figured it was only proper to take it back with us, taste it, and see what it's like before we bring the news to the king."

"Come on, Prince," the stockier one said, again. "Have a share." He held the decanter out for him.

"Raj…" Amicus' voice was barely even a murmur over Rajveer's inner turmoil. The future lay before him. He had the festival and his so-called courting of Theodora…

"I think I should go. Can't keep the lady waiting, if you know what I mean." The joke fell amiss in the room.

The Satelle shrugged, downing another gulp. They all passed the decanter from one to the next, each taking their own pull, the burn of the liquid as it went down their throats apparent on their faces. Rajveer turned to leave when laughter bounced across the table and circled around him.

Fates, save us. Rajveer swiveled back into the room; Amicus abandoned back in the hall. He reached past the Satelles, grabbed the decanter off the table, and set it to his lips.

The burn of the liquid was immediate on his lips, his tongue, his mouth, his throat. All of it was ablaze. He counted down until he would feel the inevitable dizziness, stripping him of his required duties, allowed him to feel absolute peace for but a moment. A burst of cinnamon left a delicate tingle behind in his mouth.

The Satelles cheered, the one named Clara reaching and pulling the bottle from him. His empty hand now held an echo of the guilt within his gut. If he was still having feelings, then he was still too sober. He moved to grab the bottle back and felt a strong hand on his bicep— not forceful, but enough of a warning.

Amicus tugged him away from the table. Rajveer turned his ear at his friend's words, but also kept an eye on his surroundings as the other Satelles began again with their jokes and their laughter.

"Raj, come on. We still have to get to the festival. Not only do you have an image to uphold, but also there are things we need to put in motion."

"Why? Why do *I* have to be the one to do it?" His voice was too loud in his ears. He could hear the Satelles quieting down, eager to listen in on any conversation from the prince, especially one filled with so much emotion.

Amicus pulled on Rajveer's jacket sleeve, wrenching him out of the room like a petulant child. Amicus threw him up against the stone wall and, without time for Rajveer's brain to even process it, slapped him across the face. The sting was more from shock than actual pain.

"Get it together, man!"

"What?! What do you want from me?"

"I need you to go out there and do what we planned. I need you to go see Theodora and your people, and help them celebrate this new season. Because in a few days," Amicus dropped his voice low, to a whisper crawling over Rajveer's skin, "it will be *you* they are calling king."

Rajveer wanted to fight back. He wanted to give his friend all the reasons this was absurd, all the reasons that this was absolute bullshit. For although Amicus had been a lifelong friend, their different responsibilities forced a division between them. Rajveer was unable to freely express what he truly felt.

Tugging his sleeve free from Amicus' grasp, Rajveer wavered a moment. It might have been a physical wavering too, he wasn't sure. He wanted to go back in that room with those Satelles and help them drain that bottle, to know if those gold flakes tasted any differently.

He dropped his head in defeat and sucked in a breath to sober himself against the drink, the vocatus, which now attempted to take him over. He began the trek down the hall, Amicus close on his heels, as he shoved those emotions deeper into his belly, hoping they would burn.

Festival

The sky bled red with the final hours of fiedelight as darkened rolling clouds lurked along the horizon, the fates threatening to unleash much-needed rain. The solta heat was abashedly forgotten as the chilled breeze wove around people who followed the decorated paths to the capital, ignorant of the storm brewing above.

Theodora trailed the swarm of Lumens, letting her attention shift to the tops of the trees where they peeked above the shadows from the outskirt houses, their new colorful leaves aglow with fiedelight, though a handful still paraded their greens of old. Music filtered through the clicking of heels and clomping of boots.

The opening into the square was an ambush to Theodora's senses: people shouted over laughter and music, couples danced alongside vendor carts and various game stands, scents of bread and sweets swirled around her, and clothing of bright colors mimicked the vibrant signs and banners. The Lumens' excitement was palpable. The vivid transformation of the capital was overwhelming and yet also contagious.

Theodora snaked through the lines at vendors which encircled the fountain when she recognized Amabel. The light shimmered across the rich fabric in her hair as she tilted her head back to drain a wineglass. A slight pink tinted her cheeks, and Amabel's already bright eyes lit up more when she noticed Theodora.

Amabel raced in her direction, as quickly as she could through the crowd, and threw her arms out wide to embrace her.

Theodora's suspicions were confirmed: Amabel was enjoying the king's lenience on his wine hoard, when Amabel shouted, "T!" The single letter, strung with unnecessarily added syllables, was loud in Theodora's ear. As Amabel pulled out of the embrace, she slipped on the ends of her emerald-green dress, causing her to stumble for a moment. One strap of her dress slid down her glossy shoulder.

"Enjoying yourself?" Theodora asked, reaching out to help stable Amabel. Theodora's mind worked quickly to keep her attention on her drunken friend but also the others around her, especially as the dancing shifted closer to where they were.

"Oh, fates! It's amazing, T. But, wait…" She slowed, holding onto Theodora's shoulders as she peered around her. "Where is your date? Didn't the prince ask you? You are far too beautiful in that dress to be here alone. Here, let me help you find a date." Without missing a beat, Amabel wrapped her fingers around Theodora's wrist, trying to pull her away from the dancing and toward the vendors.

Theodora chuckled, easily slipping her wrist from Amabel's grasp, which forced her friend to stop and face her. "Am, it's okay; the prince is coming. I saw you first though."

"Oh, no, don't you worry yourself with me. Go find him."

"I can't leave you right now. You are barely in a state to stand."

"Pssh." Another word far too full of syllables. "I'm here with Julian. He went to get us some food," Amabel retorted as she pointed behind herself.

It took Theodora a moment to find who she was talking about because nobody—*nobody*—called Albani anything other than Albani. Jul—nope, Albani's balding head turned, and Theodora witnessed him clunk over to them, carrying a platter of sweet bread sprinkled with soft sugar.

Amabel reached for the plate, ripping off pieces of dough. "These are so good!" She let out an audible moan. "Theodora, you need to try one." Amabel began to pull off a piece, but Theodora rocked back on her heels, away from the bomb of sugar dust, her hands up in mock submission.

"That's okay, Amabel," Theodora laughed out. "I'll eat later."

Albani placed his hand on Amabel's back, leaning into her. "Why don't you see if there is anything else you want to try?"

"Alright, but"-Amabel turned to Theodora- "you better go find your date." She swept Theodora into another embrace before flouncing away.

As Amabel left, Albani pulled klaud from his pocket and offered it to Theodora. She didn't need to ask what the payment was for. She already knew: to keep her mouth shut about what she'd witnessed yesterday. Albani was a good person, but a terrible businessman, allowing Theodora the opportunity to save both his livelihood and his reputation numerous times. At a cost, of course.

"Your confidences are starting to become expensive, Albani."

"Just take the klaud and keep your mouth shut."

"Have I spilled a secret yet?" But they both knew Theodora did well at keeping secrets.

Theodora watched as Amabel turned around the square, reading the vendor signs of the various food and drinks being offered, from frozen oranges and peppermint sticks to sausages and carnis pies. "Just be careful with her, please."

He scoffed out a laugh. "Why don't you worry about yourself and that prince of yours?"

"Watch it, Albani. Your secrets are worth more than what you think you know about me."

At his confused expression, Theodora took the opportunity to leave, gliding into the stream of people until she made her way to the fountain. She lingered by the stone tiers, little droplets of the fountain water sprinkling onto the bare skin of her neck. Here, in the center of the crowd, it was an ocean of bodies, like perpetual waves crashing about in time, swaying with the tempo changes of the music.

She was becoming claustrophobic, alone in the middle of a violent sea. Hoping to distract herself, she walked the perimeter of the fountain, looking out in every direction to find some indication of this damn prince. She was making a second pass when a young girl stepped in her path. Theodora slowed as she watched the girl and glanced around, looking for her parents or someone who was supposed to be watching her.

The girl stepped closer to the fountain, her hair dark as a raven, woven into two braids falling past her cheeks and shoulders to her middle.

Theodora reached out, stopping the girl from getting too close to the water, and dropped to her knee. "Hey, there," Theodora said as gently as she could over the vicious music and boisterous people. The girl faced her now. Her small, upturned nose looked even smaller compared to her big round eyes. They were innocent and serene. She looked expectantly at Theodora, a question furrowed in her small brows. "Where are you going?" Theodora continued, afraid to ask her about her parents.

The girl looked beyond her, apparently searching for something—hopefully some*one*, before she turned back to Theodora and held up her hand. Within her small fingers, she held a klaud.

"My gram-mama gave me this. I am promised by many, but delivered to few; I live in all, no matter who; seek me in candles, dreams, or a star; I'll bring you hope, but only seen from afar."

Theodora smirked at the riddle she was told when she was a child, the memories bubbling up again.

"A wish." Theodora whispered, and a giggle escaped past the girl's thin rosy lips. "Alright, I'm watching. Go ahead."

Staying beside her, she watched as the girl held the klaud close to her chest, her wish barely a whisper over the little token, before she threw it into the fountain water. The girl turned back to her, and Theodora asked what she'd wished for, curious about the endless possibilities of a child's imagination.

Tears shone in the girl's eyes. "For my mother to get better." She sighed and murmured, *"Alea iacta est."* Before Theodora could stop her, could find possible comforting words, the girl was gone, lost to the sea of people.

"Alea iacta est," Theodora whispered as she watched that klaud sparkle in the setting light. *The die is cast.*

Abandoned

Once close enough to the market square, Rajveer dismounted from his horse. With Amicus left behind to keep watch over his father and the castle, Jude and Miles joined him on foot while the other Satelles remained on their horses. Rajveer hadn't seen Danika yet, but he knew she would be making her appearance shortly.

The slight buzz he was blessed with from the vocatus was already starting to fade, causing Rajveer to feel even more on edge. With resolve to steel his emotions, he glanced upward, gulping down the chilled air, which had carried the thick storm clouds to the capital quickly.

When he brought his face back down, he was no longer the abandoned boy, but the Lumen prince. He accepted two wineglasses from a nearby vendor, quickly draining the first before raising the second in the air.

"My good people of Lume. The time has come when we can finally celebrate the arrival of astrum." A couple of cheers broke out among the crowd. "Change is inevitable, and nothing is promised. Let us seek forgiveness and move toward righting our past wrongs. Cheers!"

He looked out to the crowds' sprouts of glasses, wine and beer, raised into the air before their collective gulps joined his when he emptied his second glass. Jude took the empty glasses from him as Rajveer noticed Theodora standing on top of the fountain edge, her gown a brilliant shade of cobalt. He turned to the musicians, nodding for them to begin the music again, and his citizens resumed their dancing as Jude slipped two more wineglasses in his hands.

Rajveer took another sip of his wine, turning to Miles. "Theodora is over by the fountain. As I head over there to talk to her, can you keep a look out for Danika? Although Amicus said she was fine, I don't trust for something not to happen."

"Yes, sir," Miles acknowledged with a quick nod. "Although I'm sure you're overreacting."

Miles might be correct. Or maybe it was the ominous storm above them that made Rajveer feel as if the night might end terribly. He took another sip, hoping to drown the memories from his mind, because he hadn't seen clouds like this since the night of his mother's death.

Rajveer sipped again and trailed behind Miles as he parted his citizens from in their path as if they were puppets on strings. He couldn't stop the smile as he got closer to where Theodora remained perched. It was out of place to see a woman dressed in festival regalia atop the small stone wall, but when she hopped down, the act felt very much like her.

"My Lawless," Rajveer said with a mock bow as he extended a glass for her.

"My prince." Her face flushed as she accepted the glass with a small curtsy. "So, you've learned who I really am."

"Oh, I knew when we first met. Don't underestimate me. I have my own resources, too."

"Hm. Your resources are utilizing an established guard that have been with your family for how long? Not sure it qualifies as *your* resources. Respectfully, of course, sir."

"You look beautiful, by the way. Of course, with you up there, I thought you were part of the fountain."

"Well, that's good. I was attempting to blend in. Didn't want to capture too much attention."

"Unfortunately, I think you failed with that as soon as you agreed to meet with me."

"I'm not too sure. I think this gown is far brighter than your black-on-black ensemble."

"It has gold," Rajveer taunted, pointing to the gold whorls sewn into his jacket. "In case you were too distracted by my pretty face."

With a chuckle, she sipped on her glass, allowing Rajveer the opportunity to shift the conversation. "How do you typically celebrate the festival? Is it usually as grand as tonight, celebrating with the prince?"

"My blade rarely gets hired for festivals, so I would watch the passersby from my window, or occasionally study some books."

"Ah, so this is much more exciting."

"Maybe, maybe not, but this is definitely more interesting."

He watched as she sipped on her wine again, observing the way her full lips framed the glass, her slender hands wrapped around its neck. Rajveer could see how Danika, or anyone, might find her attractive, but the thought made his heart hurt as he longed for Alouette. The one who should have been accompanying him to these festivals for fiedations now.

"How do you get used to people watching you?" Theodora interrupted his wallowing thoughts.

"After a time, you just do. I mean, it makes you stand straighter, and you remain more mindful of your actions, but I guess having done it all my life, it comes naturally."

"What are you like when people aren't around?"

"For the most part, the same, because there is *always* someone. But, why such morose talk at a festival?"

One of the Satelles swooped in to provide Rajveer with another glass, stopping Theodora's response. Again, Rajveer took the opportunity to change their discussion. "Have you eaten?" he asked.

"I did, before I came here."

"Well let's go find those chocolate poppers we talked about before."

"But Amabel's attending the festival?"

"You can always find her sweets. You just need to know where to look."

Scales

The smoke of Ludi Votivi was extreme, probably the most Maddox had seen in the small den over the past few petriks. However, the Seclusians' excitement was to be expected. Although their celebrations were not as fanciful as those on the topside, they continued to find a way to make theirs worthwhile. Their underground city had continued to be ignored for far too long, and Maddox was restless to see his plan begin to unfold. But being patient was something he was very, very good at.

Maddox sipped the vocatus, staring at the other players at the table, having quickly glanced at his hand of cards. He tossed additional gems into the pile on the center of the table, increasing his bets, as he met Valix's intense stare. It was a look that Maddox had learned meant Valix hadn't the faintest idea of what Maddox held. Without giving his friend the satisfaction of guessing, Maddox sipped his liquor again. Valix called his bet, matching the gems Maddox had added, passing the decision to Mekari.

"I don't know why I even bother playing with you," Valix murmured as he nursed his glass.

"No one can hear your whining over all this excitement, Valix."

"He's just pouting because he knows he's just wasting his money when you're sitting at the table," taunted Mekari.

"No, Valix only uses half of his money on drinks, cards, and women. It's the other half he wastes."

Other Seclusians played various card games around the room: Tantrums in the Corner, Curser's Honor, Izzo Despair,

Scales of Luck, and Man and Fates. He had participated in his fair share over the fiedations and liked the varying strategy that each game brought, but he preferred Scales of Luck, playing against another person with pockets of probability.

When the wager made its way back to Maddox, it was time to reveal what he held. Five fate cards, straight pretties. Seclusians who were eating nearby and witnessed his hand hooted a holler of what should have been Maddox's celebration.

Valix angrily threw his cards down. "This is ridiculous."

"You shouldn't have expected a different result and you know it," Mekari chuckled under his breath as he gathered the cards and began shuffling them together.

"Come on, let's go again. It is logically impossible for you to win continuously."

"No, only improbable," Maddox mused, "but that's why you lose. I'll take more of your money next time. Gemma, sit in for me, please."

The twin rose from where she was perched on the back of a chair overlooking the den from the corner. "The usual, sir?"

"Yes, please," Maddox acknowledged as he stood.

Gemma took half of Maddox's winnings to the den lord, earning everyone another round of free drinks.

"What other matters could you possibly have?" Valix began to rise as well, curiosity squinting his eyes.

"None that concern you," he said haughtily as he headed for the exit.

"All your matters concern me."

"Stay, Valix. Earn back your money; maybe find a nice woman for the night. Enjoy the celebration. I'll see you tomorrow."

"You're going topside, aren't you?" Valix questioned as he trailed Maddox into the tunnels. His voice dropped to a low mumble. "Maddox, you can't keep seeing her like this. You know what our priorities are."

"I know mine. Do you know yours?"

Forgotten

The market lights blinked to life as the darkening storm clouds plunged Lume into a dusky gray. Like their time at the capital house, Theodora shadowed the prince as he stopped and socialized with people. He was the epitome of calm and collected, occasionally touching her lightly across her back or winking at her, attempting to persuade those around them of their relationship. But she detected something else. She noticed when they were closer to vendors, he would quickly consume a flute or two of wine before returning to the throngs of people. It was slight, almost indiscernible, yet Theodora saw his eyes gloss over every now and then, the way his speech pulled or slowed on certain letters.

They finally pulled away from another group of Lumens, this one talking about some possible uprising in Freta, when they made it to Amabel's cart. Amabel, of course, was nowhere in sight.

"Here you are, sir." A young boy with bright eyes handed over two chocolate poppers sprinkled with tiny red crystalline pieces along the top. Rajveer took them after tossing klaud to the boy.

Theodora accepted her piece as she and Rajveer faced each other, the music and people clinging to them, and tossed their chocolates into their mouths. It was like a mini explosion, the pops jumping along her tongue and igniting her taste buds. And maybe it was the adrenaline, or the playfulness added to the popping of sugar in her mouth, or how the wine wrapped around her middle, but she didn't want these feelings of youth to dissipate. Without a second thought, she grabbed Rajveer's hand and tugged him into the crowd. "Dance with me!"

Lumens didn't have choreographed dancing. Many, many fiedations ago, they did, and one could see those lost secrets among some of their current dances. For the most part, it was now an assortment of tangled body parts grinding and sliding across each other. For a moment, the wine had her feeling weightless, and she closed her eyes and let the passion of the music guide her to spread her arms wide and twirl to the rhythm.

Rajveer stayed nearby but remained more controlled, slightly swaying to the tempo. He talked occasionally to the Lumens around them, shaking hands with the men who wandered close enough, sometimes bending low to talk to one of the children. After a few songs full of Theodora's endless dancing, Danika approached them with another wineglass for Rajveer to guzzle. The music transitioned into a slower song, and couples folded into each other's arms. Theodora copied the movement, lifting her arms around Rajveer's neck and pulling him closer. She felt him flinch before he slid one arm around her waist, shoving his metal hand into his pocket.

"Don't worry, prince. Remember this is only a ruse." He appeared to relax with her words. A sad smile brought forth his lone dimple. She wanted to go back to how they had been moments before, ignoring their responsibilities and enjoying the festival. "You can use your other hand, Rajveer. It isn't going to bother me."

"Unfortunately, it bothers me."

Curious, Theodora questioned, "Do you want to talk about it?"

"I haven't talked to anybody about it except those who were present at the time."

She didn't respond. She let the music cover them, like a blanket they could hide under. She didn't press him to answer either, because she knew soon, she would be free from all of it anyway, banished from the capital.

"It was shortly after I met someone," Rajveer lamented quietly. "I was helping in the square…"

The fates announced their disapproval as a deafening thunderclap sounded across the sky, startling the entire festival. The Lumens glanced at one another as a deluge of rain advanced to the heart of the square.

Rajveer pulled from Theodora so quickly she was dizzied. She steadied herself and whipped her head back to Rajveer, but he was already gone. Satelles intruded into the crowd to locate him, directing him out and away as they pulled their hoods forward in anticipation of the downpour.

Danika hesitated. Theodora noticed the turmoil in her green eyes, unsure of what to do next, but they both knew that for Danika, the crown would always come first. Theodora nodded in understanding, and as Danika turned away too, she was left alone in the crowd of chaos.

Wine and adrenaline mingled in her blood. It made her feel as if she were standing on a precipice as she watched people ducking into shops and taverns, some climbing into carts if they were big enough. In a blink, the capital had cleared. It remained plastered in weeping colors. The raindrops hammering against her skin were cold, drowning her dress. Little pools of rain formed in between the cobblestones.

She was forgotten. There was only the drone of the rain pounding the ground and the roaring in her ears. She wanted to be an ordinary person. One with a family, one where she felt loved. Was it such a fanciful idea?

Gathering her skirts, she turned in the direction of her home, her clothing drenched and heavy in her arms. As her boots splashed in puddles riddling the street, she passed stragglers who hadn't yet found cover, until she stopped short.

Maddox stood with hands tucked deep into his trouser pockets. He wore no jacket, but he had a brilliant blue vest that somehow matched her dress perfectly. The white of his shirt clung to his muscled skin. Her heart cracked when she saw him, and relief washed over her, almost as quickly as the storm.

She raced forward and he bent down to meet her, cradling her body against his, hands wrapped around her chest and burrowed into her hair.

He pulled away, the warmth of his body vanishing as he placed his hands on her cheeks, his dark eyes deeply searching hers.

"I'll always find you, charm."

Theodora had no words. They stuck in her throat, swirled with too much emotion and wine.

So, she kissed him.

She grabbed his shirt, pulled him to her, and kissed him.

But it was nothing like the night in the study. There was no hesitation. She filled the kiss with the slow burn she had felt for him long before she realized the spark had caught.

He broke it off, far too soon. "Let's get you home."

She felt the disappointment spread across her body and was planning to argue with him until she noticed the boyish grin that smirked across his face.

Unsteady

Arriving at the castle, Rajveer immediately dismissed his men. Some went to ensure the care of the horses, while others helped patrol the perimeter against the threat of the storm. A handful clustered at the front entryway, staring at the onslaught of rain. Big, bulking men ran off like street rats with the threat of water.

His head had started to pound during their ride back as the buzz from his copious amounts of festival wine began to dissipate. He marched the halls back to the Satelle sitting room, and as he expected, the former occupants had never finished the bottle of vocatus. There wasn't much left, maybe three fingers worth, but it was something.

Walking to the doorway, he looked down the hall and spotted some Satelles, none of which were headed in his direction. He tucked himself back in the room and took advantage of his isolation to rip the stopper from the bottle and drain it of its contents. The cinnamon burned his throat on the way down; the fire he craved.

But it wasn't enough. It was never enough.

With the Satelles busy with their own tasks, he hoped his absence would remain unnoticed for a bit longer. Abandoning the now empty bottle on the table, he proceeded to the stairs leading to the underground levels of deserted rooms, old prison cells, and ignored tombs.

His boots shuffled on the stone steps as he descended the darkened stairwell. The rise of must and decay had always been enough to make him nauseous on a sober day, and now with a mix of vocatus liquor and numerous glasses of wine, his stomach turned painfully in disgust. He was heading down blindly, holding onto a sliver of soberness and his memory to get him to the bottom.

"Why would we need light to walk down the stairs," he mumbled sarcastically as his hand searched along the wall for the recently installed pulled chain at the bottom. Dampness from either sweat or rain clung to his forehead, which he smeared with his hand before pulling the chain, illuminating the hallway in a dull glow. The stones were warmed by the light, shadows creeping along the edges. Dark wooden doors were erected on either side, alternating back and forth. As he advanced further down the hall, his stomach rolled about, compelling him to stop again. He pressed his head against the cool stone and forced deep breaths to fill his lungs.

He had a reason to be down here, and by the fates, he was going to do it.

It was slow progress, but he shambled down the hall, passing over ten yards of repetitive door after door until the walls finally spilled outward into a giant circular room known as the Klauduisz tomb. There were other tombs, discarded within the maze of doors and halls, but here the walls were decorated with darkened wood bearing metallic rectangles with etched names of the ones held within. Of the nine places, five of them were left unmarked, although three had already been claimed: ones for not only his father and himself, but also his mother.

In the center of the room was a stone sarcophagus, her features etched into the material as if she were still here, merely taking a rest, and would awaken at any moment. They hadn't yet placed her into her own tomb, for his father had been unwilling to part with her, even well after her passing.

Rajveer stopped at the opening and took in the space, muted from the distant light, cobwebs and dust invading from every corner. He slowly entered, bringing himself closer to his mother, staring at her face and the way her nose turned slightly to the right near the bottom. Her lips were set in a small smile, and somehow the stone made them appear soft and radiant, although it failed to capture the coloring of her eyes or the slight pink of her cheeks.

With a slight shift, he stripped his jacket off before sliding down next to the altar, the movement causing his vision to spin. He let his head fall back against the stone and breathed through the nausea and perpetual feeling of movement, even though he was sure he was sitting completely still. He alternated between keeping his eyes closed and open, not sure which one helped.

He forced his fingers into a chip where the floor and altar met, allowing him to lift the floor tile enough to slide his other hand under and retrieve the wine bottle stored inside. It was half gone, but it was a start. He tugged the stopper out, having to use his metallic hand for the strength to remove it.

After returning the tile, he rummaged through his discarded jacket, searching for the inside pocket. He located his pack of cigarettes and lighter. He struck the light and sucked in a breath of the cigarette. The smooth, rich flavor filled his mouth, and a hint of licorice sat at the back of his throat before he let it out in a swirl. He brought in another drag.

He exhaled and broke down, cradling his head in his hands, which still held onto the cigarette and neck of the bottle, as if they were his only lifelines. "I'm so sorry, Mother. I left her. I left her in the rain like she was some Lawless trash. What if we were courting?"

His voice was hollow in the stone room. "I don't know what the right answer is anymore. I don't know what I'm supposed to do. I know what everyone else wants from me. I know father wants me to marry, to carry on the Klauduisz name. Do I do right by him? But what about the Lumens? Am I to abandon them?"

He brought the cigarette to his mouth once more, taking in its roundness. He let it fill him, sober him, ground him.

"And what about Seclus? Amicus thinks it's best to take the throne, to try to get rid of Seclus completely and take back control of our world. But what about the future? What about *after* me?

"I've been so stupid, Mother. I fucked up. I know Theodora isn't going to replace Alouette, no one ever could, but she makes me feel less lonely. She makes me feel alive, like maybe I could take on all these pressures placed on me. She reminds me what unconditional love feels like. And I know that's weird, because I don't love her, not like that, but…"

A groan escaped, reverberating with the stress and insecurities deep within his core. Rajveer didn't know the right answer for himself, but he was never meant to be anything else in this life. His one love left him the night before their wedding. His mother died well before she could see him claim the throne. His father didn't give two fucks if he lived, or died, or married. He knew Amicus was right. His father would find any excuse for why he couldn't take the crown, letting Lume fall further into oblivion.

Rajveer would claim his title and start writing the stories of the one thing he was born to do, raised, and reared to be.

Hope

They made it through the rain, banging up the steps, water dripping from the two of them. They left boot marks and handprints throughout the stairwell. The storm had taken out the building's electricity, and with the never-ending thunder, it appeared it wouldn't be returning any time soon.

Barely visible in the dimness, Theodora fumbled with the lock on the door. When she finally freed it from the clasp, only a muted light from the stormy skies snuck into the room.

Theodora searched through her cabinets for candles and a lighter while Maddox hovered nearby. She knew they were in one of the lower cabinets, most likely coated in dust, but couldn't recall which. The smell of damp clothing began to fill the room as the harsh winds of the fates hammered outside. Finally locating them, she set the candles throughout the washroom and on the tables in the living space, carefully moving scattered papers to avoid the drops that fell from her hair and body.

Once the candles were lit, the little spheres of light from each candle stretched out, and she stole a glance to Maddox. He felt like a dream or part of her imagination, fleeting and magical, as if she were to close her eyes, he might *actually* disappear into the shadows.

He met her gaze and closed the distance between them. He brought both hands carefully to her face as she searched his, but for what, she didn't know yet. Maybe the answer of what was to come, of what this was supposed to be.

When he spoke, his voice was low and gruff. "As much as this dress is gorgeous on you, I think it would be better on the floor."

"With me in it?"

"You're a pain in my ass."

"I'm glad I could be of service." Theodora stepped away, out of his grasp, and continued walking backward to the washroom, never removing her eyes from his.

Maddox followed her and stood in the doorway as she lowered herself to the edge of the tub. As she unlaced her boots, Maddox removed his shoes. The tension between them filled the little room to the brim with wants and possibilities. She reached up her dress to unclip her soaked tights, water dripping from them as she pulled her legs free. She began unclasping the front of her dress as he undid his own buttons. Their gazes never fell, watching each other remove their layers of heavy fabric, like armor strategically placed around their hearts.

Before she let the dress fall, she hesitated. The unasked question hung in a balance between them.

Maddox removed his shirt, and it was the first time she had witnessed his bare skin, save the moment in the alley with physical wounds and blood. She admired the way his skin glistened with water droplets as they fell from his hair. Scars scattered along his chest and abs, the wound from a few days ago still red and angry. Her eyes crept down his body, snagging on where the fabric of his trousers sat low on his hips.

Maddox crossed his arms and gave her that damn smirk. She fought the urge to tuck down her face, to keep the blush from creeping up from her chest. She shimmied through the fabric of her dress, letting it fall into a pool on the floor as she stepped out of it.

She stood before him in only her intimates, corset, and the dagger at her thigh. She walked forward slowly, reaching up to touch the front of his hair, the few strands that fell unceremoniously onto his forehead. A water droplet had formed at the end, and when she touched it, the coolness slipped down her finger into her palm. Maddox reached up with both of his hands and pushed his hair back with his fingers.

He wiped his hands down his face, and when he looked at her again... fates, his stare had turned fierce. She sucked in a breath when he knelt before her, the act like what he had done earlier that day. The memory of his fingers touching her skin had her quivering in anticipation.

Her dagger was the last piece, the last line of defense to keep her heart safe.

His rough fingers walked along the skin of her thigh as he unbuckled the holster, the metal and leather clinking on the floor tile when Maddox put it down. With the shadows from the candlelight, and maybe the secrets of their already shared kiss, everything was more intensified. She threw her head back in expectation, a tightness forming in her center. Every place where their skin met felt like a small fire, and the only way to snuff it out was by pressing harder.

Maddox stopped abruptly, and it left her feeling dizzy. She gulped down air as if she had forgotten to breathe, and he rose in front of her. She raised her head to follow his, the full height of him towering over her. The calluses on his hands scraped along her cheeks, and she searched the depth of his eyes. The unspoken question laid between them.

But Theodora failed to speak. For Maddox had destroyed every line of defense she had ever created to ensure she never had this feeling. She nodded her head as she forced the syllables to scrape along her vocal cords, the bare affirmative mingling with their shared breaths.

It was the only thing he needed.

Whatever monster he had held at bay was unleashed. He surged forward, his mouth meeting hers with desire that made her ache. His passion ignited the embers that had laid in wait, the sparks dancing to take hold and catch into a wildfire. Their tongues mingled and she tasted beer, beer he must have consumed at the festival. When Maddox attempted to move away again, Theodora sucked on his bottom lip, acting on the desire for him to stay close. He bent forward and grabbed her thighs, forcing her into his arms. Her hands ravaged his body, her fingertips traveling along his scars. The rise and falls of his muscles. The barely discerned stubble along his chin. His thick, coarse hair as it slipped through her fingers.

It was new and foreign, but somehow still deep-rooted.

Theodora felt as Maddox's legs struck the bed. The cold sheets raced up her warmed skin as he slowly lowered her. He held himself above her with his elbow and looked into her eyes once more. His lips quirked up to the side. It brought light and joy to her thundering heart.

He leaned down, pressing his forehead against hers. She took a moment to breathe in his scent, let it fill her lungs. He smelled of cool air and fresh night, her freedom and escape. And fates be damned if she wasn't going to take it. The fire they had started crept from her center and burned within her. It blazed anew and there was nothing to be done now to stop it.

∴ ∵ ∴

Maddox collapsed onto the bed next to her as they both attempted to calm the thrumming of their hearts. Their broken gasps for air filled the room, his arm against hers sticky with sweat. She pulled away and turned on her side facing him, kissing his cheek and neck before placing her head on his shoulder. He leaned forward and grabbed the blanket from the edge of the bed, throwing it over their naked bodies.

Physically she was exhausted, but the idea of Maddox spending the night with her and the unchecked task of assassinating the king both stalked her mind. She closed her eyes and focused on his breathing, the winds outside a muffled harmony to her own breath.

She startled when Maddox spoke, his deep rumble filling her ear. "Do you remember that day in town when we were children? You were hiding in one of the alleys…"

She interrupted him by saying yes, attempting to save him from explaining the details.

He turned to face her, forcing her to move her head to a pillow so their eyes could meet. He opened his mouth to speak, but the words failed him. Theodora searched his dark eyes, trying to find those words for him. But as dark as they were, the emotions he held within were even more hidden. She took a leap, hoping that she was not alone in what she felt, the warmth of those flames.

She placed a hand on his cheek. "I've carried your heart with me since that day."

"I will always find you, Theodora."

And there were no other words. As much as she wanted them, she couldn't begin to find them for him. So, she showed him. Showed him the only way she knew how. Again, and again, until the monsters that circled her thoughts were put back onto their leash, and she finally found peace in sleep.

Memories

The cobblestones were wet and flooded from the aftermath of the storm. Maddox walked the capital streets with nothing but damp clothes covering his skin, using the advantages of limited Satelles on patrol and failed electricity to keep him hidden.

The scent of the newly fallen rain mixed with dirt and stone. Finding the path out of the capital, he set off for Seclus alone. He thought about telling Valix or the Gems where he was headed, but wanted to avoid an earful or scornful looks because they knew the trip was merely for selfish reasons.

He had listened to Theodora as she fell asleep. When she tucked her head into the crook of his arm, she had drifted quickly into a peaceful slumber. It was bittersweet for him to witness, her ability to be so at ease with him while he was unable to sleep next to anyone. Then again, he didn't know the last time he was ever able to simply fall asleep. It was a process and an ordeal he dreaded with the close of each day. As much as he enjoyed this night with Theodora, there was no way he would be able to fall asleep any time soon.

He didn't know why he'd recently decided to pursue her, especially after so much time being distant. Distance was safe, acceptable. And he knew he was taking a gamble with each passing day.

Maddox assumed he had feelings for her, but he was incapable of deciphering them. Maybe he cared for her? He wasn't sure. Caring seemed inadequate to actively describe what he was experiencing. Instead, he repeated those words to her over and over. *I will always find you.* Were they a promise?

An oath he didn't realize he was making intertwined with the phrase?

But his timing had been short of callous. With this challenging task looming over them, he felt himself spending more time focused on her rather than the more pressing matters at hand. Rubbing his fingers along his brow, he attempted to remove the thoughts that swarmed him. This was why he preferred numbers and calculations to people.

He concentrated on the sounds around him: the insects, the swish of the dried grass as it swayed in the breeze, the crunch of the ground under his boots. Focusing on the journey, one he had taken many, many times, but rarely on foot.

Except one time when Valix was piss-poor drunk and they ended up commandeering some Lumen horses. But as one memory surfaced, more memories exposed themselves. Belly up, unprotected, and ready for him to recall. If dragging up the past was his punishment for taking the long way home, fuck it.

Maddox pulled the pocket watch from his vest. As he opened the face and began making the appropriate turns and clicks, he reminded himself of this final task. After this assassination, he could finally return home. As the portal opened before him, he saw Valix on the other side of the ripple. Maddox walked through to join him.

"Maddox! I was wondering if you were going to make it back." Valix rose from one of the boulders tucked along the side of the tunnel entrance of Seclus.

"Are you a mother hen now? What are you doing up here? No woman wanted to share their bed with you?"

"Oh, I had plenty, don't you worry. The Seclusians had their own celebration tonight." Valix stroked his chin as if recalling the night. "I think there was a man involved as well, but I can't be sure."

"Did the retaining wall hold well enough?"

"Why wouldn't it? She held up fine. No one had any doubts or concerns."

"Good."

They both strolled past the main boulders into the entrance tunnel, and Maddox's eyes bounced back and forth to either side, checking where the retaining walls were tucked into the rock.

"What's on your mind, Maddox?"

It was the one thing that bothered him about his second, his ability to read people and sense when things were amiss. Although it was an advantage when being used *for* him, he didn't enjoy it being used *on* him.

"Nothing."

"Did you see Theodora at all when you were up there?"

"No."

"No? That's it. Just, no? For some reason I'm not buying it."

Maddox shrugged, "Suit yourself."

"Then what were you doing up there?"

"Don't worry about it, Valix. You worry about the tasks I've given you and make sure they are going to be taken care of."

"They're already done, sir."

"As they should be. You don't have this position due to your ability to ask questions."

"Yes, sir."

And as easily as Theodora had taken down the armor around his heart, he put it back into place. But with each rise and fall of his armor, Maddox felt it growing weaker.

Undeniable

Peace. It was an expectant word. The possibility wrapped up in its heavy vowels. Happiness and hope swirled with a sense of calming, leaving one so at ease in the world.

Theodora had spent so much time without it. After losing her parents, there had been nothing left to ground her to Lume. She felt no permanent connection to it, aside from earning enough klaud so she could find a way to leave, to escape her past etched into the capital buildings.

Danika had promised this peace, but her solutions only left Theodora feeling caught within the king's grasp, trapped within the castle walls or the capital limits. And eventually, she knew Danika would give up trying to help Theodora find it.

Maddox was her grounding. He did not attempt to *save* her from this world, but rather stood with her, side-by-side. He showed her a place that made her feel safe, albeit unorthodox. They gripped each other and refused to let the other falter. Surrounded by Lawless who were tired of the coercion to follow broken traditions, surrounded by Seclusians who saw how novelty can spark an idea, they were the innovators, the tinkers, the dreamers.

But without change, what were they even doing? Their meaningless lives would turn to memory. Time would inevitably erase those, too, until they were forgotten once more.

Theodora awoke with a start. The blanket slipped from her naked body as she sat up and stared into the room. The morning light crept across the floorboards as she attempted to sort through her dream. The silence in the room was suffocating compared to the liveliness of the Lumens outside, full of the promises of the second day of the Astrum Festival, to make things anew after the destruction from the evening before.

She turned to the empty place next to her on the mattress. It would be a lie to say that she was surprised to see Maddox had slipped away under the cover of darkness. It was undeniable that there was something between them, but she feared what it could mean. She assumed Maddox felt the same as she did. At least, that's what she told her fleeting heart.

The chilled air of astrum wasted no time in reminding Lume of its arrival. She tugged the blanket around herself, stumbling to the washroom to deal with her discarded damp garments. But she noticed Maddox had already taken care of them, her clothing hanging limply off hooks in the corner. She felt herself smile as her lips turned upward slightly.

She turned back to the kitchen area to put a kettle on, and rummaged for tea leaves and rose petals as the water heated. Her eyes caught on the astrum flowers Maddox had brought her yesterday. Her smile grew as she poured the hot water, watching the clear liquid turn amber. Maddox had always acted as if she was some type of enigma, one he attempted to solve. But what he didn't realize was Theodora found him to be a puzzle, too.

Companion

Rajveer walked the streets, weaving around various people. He took another bite of carnis pie, a buttery crust surrounding beef and root vegetables with a perfect amount of herbs. It was comfort for his pounding head.

The Lumen capital was readying for a true celebration tonight. The excitement was as fragrant as the spice from the rosemary and marjoram in his mouth. Although the fates stole the previous night away from them, his citizens would be damned to let it happen again.

He and his Satelles were starting their seventh turn around the market square when Amicus slowed next to him. "We can't keep walking circles all afternoon."

"And why not?"

"Because… people are going to find it suspicious if they haven't already. And I need to get back to the king."

Amicus was right, though he didn't want to admit it. Rajveer hoped Theodora would have made it to the market by now. But after walking circles and even getting food, no one had seen or heard from her. Although the tart had helped to curb some of his unease, he could sense his Satelles were getting agitated.

"Well, I'm not just dicking around, Amicus. There is a reason for why we're here."

"I know. You told me. And like I said it's a stupid reason. Dicking around would be far more entertaining right now." Exacerbated, Amicus rubbed his fingers along his forehead as he continued his search of the crowds.

"Why would you think it's a stupid reason? You don't think that I should apologize for having not only myself, but also my whole squad of Satelles, leave her in the rain?"

"It's rain!"

"Exactly!"

"I don't understand your obsession with her. Is she helping us? Yes." He dropped his voice low. "But she is still Lawless. In a couple days when all this is done, she will be gone, off to whatever she has planned. No matter what happens between the two of you from now until then, she *is* leaving."

"It's nothing like that."

"Rajveer, seriously. Listen to yourself. This is all a lie. A ruse. Some wild and fanciful story. Just focus on the next few days and then we can figure out what to do next."

But Rajveer didn't want to only sit and wait. Wasn't that what he had always done? He contemplated getting Danika, but was still hesitant to use her when it involved Theodora. However, after they had made their eighth pass around the market, he knew he couldn't argue with himself anymore.

"Fetch me Miles," Rajveer directed to Amicus, who, with a whistle, waved over the young Satelle. Miles' ginger hair was easily spotted across the square, and the tips of his ears blushed pink from the attention the exchange brought.

As Miles advanced in their direction, Rajveer felt a tug on his jacket. Looking down, he saw a young boy, probably in his sixth or seventh fiedation. Rajveer knelt to the boy's level and saw streaks of dried tears and puffy eyes, which were baby blue. Rajveer's heart stuttered as he put a hand on the boy's shoulder.

"Can I help you?"

The boy choked back a noise before whispering, "I've lost my mother."

"Oh! Not to worry, lad. My astounding Satelles can help you find anyone." Miles had met up with them, and he and Amicus both shifted uncomfortably. "Was she in the market with you just now?"

"No. I mean, she died." The boy's eyes welled up with tears. Rajveer pulled the boy to his chest and did not speak for a moment. As the boy sobbed into his shoulder, he saw an older man with a strong resemblance to the child. His mousy hair was tousled as he removed the hat from his head. When their eyes locked, the man began to proceed forward.

Rajveer continued to hug the boy, afraid to let him go. The man looked down at his worn leather boots before meeting Rajveer's eyes once more.

"I'm sorry he came over to interrupt, my prince, but he was very adamant about it and rambled on and on about how you would understand. His *own* father 'wouldn't understand, but the prince would.'" He let out a nervous chuckle.

"How did she pass? If you don't mind me asking."

"The same way as the queen."

The healers had said it was a disease, some cancer that had taken over. Almost as soon as Rajveer's mother had started to complain about body aches and nausea and endless coughing, it had already taken control of the rest of her body. *Every day she gives us is a blessing from the fates.* That's what they had told him, as if that would ease the wound that had pierced his heart. The only remedy Rajveer had ever found useful was at the bottom of a bottle. But it was never permanent, and clearly wasn't one he could share with a child.

He pulled the boy away from him, and the eyes that met his were brimmed with red, now even more swollen from the fresh tears falling down his face.

"You were right, I do understand," Rajveer said, willing his voice to stay strong. "I'm sorry."

The boy sniffled. "Why does everyone say that?"

"We're sorry that you must go through this. No one should have to, but everyone ultimately does."

The boy's bright eyes met his again, and it was clear that he was analyzing and tearing apart each word, sorting and making definitions of his new life as a motherless son.

"There are no words that will make this feel better. There is nothing I can say that will make the pain go away or erase the hurt you feel."

Even though the situation was difficult, Rajveer was proud of himself for finding the necessary words to comfort the child. Maybe he *could* do this.

"Will it ever go away?"

Rajveer took a moment to swallow his own feelings. The knot in his throat hardened, and it took every ounce of energy to get it down.

"No." He couldn't lie to the boy. He had an entire life ahead of him. And if Rajveer spilled this one lie, the boy would only curse the prince's name until the end of his life. "I wish it did, he went on, "but it doesn't. You grow. You learn to adjust and live with it. But the pain never goes away. It's a wound that never heals, and it will reopen again and again."

"But how do I learn?"

"Time. Time is the only way to learn. And it's okay to talk to your father about it. He understands it as much as you do. You both lost the same person, and together you can find comfort in each other."

"Is that what you did?"

"Yes." It was the one lie Rajveer could gamble on speaking, because soon, it wouldn't matter either way.

The boy nodded as he looked at his fingers ringed together, gathering the courage to take in a deep breath. The child turned back to his father and told him he was ready.

The father nodded as well and thanked Rajveer, as if he had done anything, before sliding his hat back on and turning into the crowd with his son.

Rajveer noticed the Lumens around him, a few having stopped walking to witness the exchange. Some turned and continued their ways when the father and son left. Rajveer spotted a few who lingered: a plump woman dabbed her understanding eyes with a handkerchief, a man with an overly large top hat gave a sympathetic nod, and a slender black cat watched him.

Well, maybe the cat hadn't *actually* stopped to witness the conversation. The feline licked its lips, and Rajveer pulled out another carnis pie, offering a small piece to the cat before it trotted along toward the vendors. Rajveer took a bite of the pie himself, letting the moment settle around him. The familiarity of the food provided a comforting buffer for his heart. Rajveer turned to the warm eyes of Miles, and his Satelles' gaze pained him. He could see the boy in Miles, see his unanswered questions about why he was continuing down this path.

"Can you have a message delivered to Theodora?" Rajveer asked. "I don't know where she resides, or if she is even there, but I heard Danika knows. Although if Danika had reported this morning, we probably wouldn't have wasted half the day searching."

"Raj." Amicus' warning was clear in his voice.

Rajveer ignored him and continued his instruction to Miles. "Find Theodora and tell her I need to speak with her. Tonight. Send her to the castle."

"Absolutely, sir." Miles disappeared back into the crowd.

"Alright." Rajveer shifted his body to face Amicus as he threw the last of the savory pie into his mouth. Without fully swallowing the food, he mumbled, "Lead the way."

With a shake of his head, Amicus turned and headed for the stables to gather their horses. If only Rajveer could shake off his own feelings as well as Amicus shook off his antics.

∴ ∵ ∴

Rajveer dropped his head back into the water of the lake, exhausted. The first night of the Astrum Festival was overwhelming in its celebrations, and although half the night had been ruined, the emotions in the aftermath left him feeling hollow.

He'd enjoyed that the second night existed for the spectacle. The vendors, of course, took advantage of it to sell their valuables, but performers, such as musicians and jugglers and acrobats, counted down the fiedelight until the true entertainment could be displayed with the cover of darkness.

The light was now a fading purple in the sky, making the surface of the water appear like an inky black. He let his mostly naked body float, his ears filling with the water, muffling the world. The occasional swish of water on sand echoed its way up to him. In this moment, his brain ceased to rumble through thoughts and feelings.

He occasionally enjoyed nightly dips, but typically stayed ashore. Even during the solta festival, when the teal water would be crowded, he remained walking the sands, mingling with the other Lumens.

Rajveer took another deep breath when he heard a splash nearby. He jerked his body upright, treading to stay afloat as he scanned the water. The worst creature in here was probably a water snake, and he didn't want to be bothered with it. His skimming of the coastline stopped when he located a silhouette, the caste lights bright beyond it. Curious about who would be out here, he swam in the figure's direction until he recognized it was Theodora.

The disdain on her face was as clear as his reflection in the water. Her wavy hair was pulled in a low ponytail, draping over her left shoulder. She stood with a rock clenched in her fist.

"Do I need to throw the second one?"

He stopped approaching when the water became shallow enough that he would be forced to stand. "Although I wanted to speak with you, I wasn't expecting you to arrive while I was out here."

"What do you need, Prince?" The emotion that poisoned her words was a slap across his face.

"I really didn't want to do this across a vast distance, Theodora."

"Are you asking me to join you?"

He contemplated the idea. Ruse or not, he didn't want Theodora to think this was something else. "I wasn't trying… I wanted to speak to you properly."

Seemingly without a second thought, she unbuttoned her trousers, pulling them down as she slipped off her boots, before she tugged her shirt over her head. Rajveer turned away from her, fearful of crossing this imaginary line he had drawn.

He felt the water ripple around him and found she'd sunk low into the lake, her white undergarments a faint glow under the water.

"Yes, Prince?"

"Please stop calling me that." His voice was gruff, but he tried to keep the small rise of anger out of it.

"What? Prince? It's what you are."

"I don't need that reminder from you. *You* don't remind me of that."

"I've called you it before."

"Yes, dripping with sarcasm." But his irritation was showing and, again, this conversation was going nowhere he wanted it to. She said nothing, staring at him to continue. "I wanted to apologize for how I acted last night."

"Stop."

"No, Theodora, I need..."

"Rajveer." The use of his name interrupted his thoughts. "You left me in the rain. It's just water." Even though she shrugged as if to show the nonchalance she felt, he could see the small sadness that blinked into her eyes. "It wasn't my first time being abandoned, and I'm sure it wouldn't be my last."

"But I wanted to be better than that."

"Everyone makes mistakes. Fates, even I do on the rare occasion."

"I'm sure it is rare, indeed."

"So, is this how you woo women? Make them stand awkwardly in the shallows of the lake?" He let out a chuckle before she continued, "Come on, before the flares start. I'll race you to the boulder over there."

Theodora started for where a large rock jutted out into the middle of the lake, as if trying to reach its separated brethren on the other side, when Rajveer launched forward to follow. These small moments made him appreciate Theodora's company. He found himself forgetting his royal title and instead enjoying life.

Rajveer slid past her, using his height to touch the sandy floor as an advantage. He turned to face her, crossing his arms over his chest, and leaning back into the boulder as she stopped short. She rolled her eyes at him as she treaded in place.

"I thought maybe you were letting me win, but the irritation on your face proves otherwise," he taunted.

"Hm, but we didn't discuss the prize for the winner." She inched closer to him, and Rajveer tried to press himself into the rock, wishing he could fall through it. "Erm, I thought the flares were the prize?" But his voice wobbled. The lines between deception and reality were again eroding, ones he didn't want to see washed away. She was so close now, and she gripped his shoulder, allowing herself to rest from swimming.

A large flare ignited the sky, the boom echoing around them. He looked up and saw the cascade of glittering lights as they fell back to the world. He let out a sigh of relief at the interruption, but stiffened upon realizing Theodora hadn't turned away from him.

She was so close. Too close, because all he desired in that moment was Alouette, that *they* could be out here witnessing the flares together.

"Theodora, I can't do this," he whispered as the flares continued behind her.

"Do what? You fuck all the other ones he thinks you might pick to take the throne with you."

He let out a sigh, pinching his brows together with his fingers. Only his trusted Satelles knew his secret., Could he trust Theodora with this information as well? "Taking them to bed is a lie. I pay them to leave and to spread gossip of our relations, so my father and the Lumens don't lose hope."

"Wow." She let out a breath but didn't move closer. "So, it's been how long?"

"Long."

"But why? Why all the false grandeur?"

"I have," he paused, searching for the right words, "a lot of expectations. Unfortunately, I can't love anyone else. I mean, I appreciate everything you're doing for both Lume and myself, but I can never love you. That's not to say I'm not enjoying your company though. You make me forget who I am sometimes, and it's nice to get away from this desperate reality."

She stared at him, and he was afraid that he might have ruined this, possibly pissed off the person who would help him gain access to power. He wasn't sure if he had stopped breathing, the flares lighting the sky in brilliant shades of pink and green. "It's a good thing I have no intentions of fucking you either, Raj." She jerked her chin upwards, behind him. He made to turn, but she stopped him. "Fates, you're definitely not Lawless. Don't make it obvious you're looking." Although annoyance filled her words, her eyes lit up smugly at his incompetence.

He squinted in the darkness to the balcony, where he made out the shape of a person. The flares provided him the occasional light to help him find his father's nightgown, white in the dark. He faced Theodora and realized it was all part of their ruse again. Relief flooded through his body once more.

Rajveer placed his arms on her waist and forced her to turn around, her back to his chest. They stood in silence, letting the weight of the world fall off their shoulders into the depths of the water. He wrapped his arms loosely around her neck, placing his head on top of hers. She leaned back into him, letting her hands rest along his arms.

Her hand touched the place where one of the metal rods entered his skin, and he stiffened at the touch. She must have noticed his tension, because she said, "Do you want to talk about it?"

"No," he said quickly. Instinctually.

She squeezed his arm. "How awkward it must be to fear love itself, but so desperately want companionship."

Rajveer appreciated how quickly she had deciphered the situation. All he wanted was to feel less alone. With Theodora, there were no expectations, no demand for him to be the *prince* she so annoyingly teased him about. He could just

be Rajveer. And maybe one day, when all of this is over...
"What do you plan to do when your task is done?"
Another explosion. A swirl of golds and greens.
"Retire."
He could hear the smirk in the word. He chuckled.
"Where are you headed? Picked out a place yet?"
"I have. I've looked at maps to try to figure out the best place to go. I debated Perdit, but I thought it would be too close to Seclus. A couple astrums ago, I found an abandoned cottage along the river off on the trade route to Nemaaer. I'm only hoping I can get enough klaud after this to never have to work another day. It's small and needs some serious work, but it's freedom."

They were silent, because it was another thing they shared, the desire for a small bite of freedom from forced expectations. And soon? They would both be able to taste it.

"Rajveer," she paused. "I care about you, too. I don't know how or why, but I do. This whole evening might have been part of our scam for the king and for Lume. And I know our entire start of this... relationship...is based on lies and deception, but that story about my little cottage is entirely true."

"Thank you," he whispered. He had no other words for her. She was going to save him, without even realizing it.

Dance

The views of the capital from the rooftops were a rare sight for most. Maddox had found these treasured spots after he witnessed Theodora scaling the walls in their youth. Curious about where she was going, he'd followed her, learning quickly she had no destination but these wooden shingles. With time, he discovered, as expected, that some roofs provided better scenes, as some had higher angles and others were better maintained and sturdier.

The one he stood on now he assumed was Theodora's favorite, as he had found her here more often. From this vantage point, the grand fountain with the bustle of Lumens could be seen as well as the castle winking further on the horizon. He had to admit, she'd found a captivating view.

The spices of foods mingled with the plumes of smoke as the laughter of children echoed off the brick walls. The excitement drummed anew, full of missed opportunities. Flares sparkled in the night sky from his unhindered viewpoint, lighting up their world for a moment. Petram was forced to the background of the flares' stage.

Maddox leaned against a chimney stack and watched as Theodora climbed up onto the rooftop, seeming to be tired or even exhausted. When her attention finally found him, she hesitated a moment before pulling herself onto the shingles. As she got closer, he noticed the shoulders of her tunic were damp from where her wet hair rested.

"Are they done?" she asked as she spun away from him to look at the darkened sky, the flares having stopped.

"No. I overheard a couple talking about prepping for a second round. I guess they want to make up for yesterday."

She nodded before scanning the sky and the Lumens below. Adequately satisfied with whatever she saw, she backed away from the edge and leaned against the chimney next to him.

"Was it not good?" He glanced in her direction.

"Was what not good?"

"You're wet, exhausted, and appear to be in an irritable mood. I assumed the prince didn't live up to the gossip."

"He didn't bed me, Maddox. It was nothing immoral, only swimming."

"Did you drown him?"

She full on glared at him now. "I said it was nothing immoral."

He raised an eyebrow as if to emphasize his question.

"Fine, I didn't drown him either."

"He's just a bad kisser then?"

"You're awful, you know that?" She swatted a hand in his direction, but a small smile formed across her mouth. It was all he needed to see to know he had won. "He wanted to apologize for last night. When I showed up, the king was there, so we got close enough for assumptions to be made."

"I see. So then why do you seem so tired?"

"There's a lot going on right now. Being called on by the prince meant riding my horse—"

"Down River."

Her small smile came back. "I forgot you knew that. Yes, it meant riding Down River, finding the prince, swimming, and returning home, merely to push through Lumens and climb to a roof. How did you know I would be up here anyway?"

"Call it intuition."

"That would imply you have some feelings." She bumped into his shoulder before continuing, "You do trust me, right?"

"Of course."

"So, you know nothing—"

"I know, charm." He was torn over whether to continue. He had witnessed lovers before, and knew all the expected responses. There were normal conversations they should have, but he refused to do any of that with her.

"Rajveer wants to go over the final plans. He mentioned meeting at the Digere again, but I thought it would be best if it were more public. One of their minions will be sent out to tell us what they decide."

"What are we going to do in the meantime?"

"Wait. Maddox without a plan?"

"I have a plan. I wanted to know if yours matched the correct plan."

Theodora's laugh escaped quickly, more of a loud sound than an actual laugh. "*My* plan is absolutely nothing. I'll rest. Maybe drink some tea." She fought with one of her loose strands of hair, attempting to force it back into its leather cord. "What's your plan?"

"Seclus."

"As always."

He huffed out a sigh. "Someone has to remain in control."

"What do you think will happen with Seclus?'

"Right now, that remains to be seen, but I am curious to hear Rajveer's plans. Not only for Seclus, but also for Lume."

A flare sounded, continuing the festival as the sphere of brilliant colors expanded across the night sky. This time, several of them shot upwards, smoke filling the air with metal salts and perchlorate. With each thunderous noise, Maddox lowered his guard once more. "It's been a while since you were up here, Theodora."

"It's one of the best spots to witness the flares."

"Is it? Better than the view from the lake?"

"Absolutely."

"You know when we were younger, I used to sit up on a different rooftop. The one over there." Maddox reached around her and pointed off to her right before continuing, "I would sit and look at Petram night after night, trying to figure out what my future would hold. What my purpose was. What I was meant to do."

"Did you ever figure it out?"

"Keep looking up."

"I am," she whispered.

He looked down to meet her gaze, but with her, he was always looking down. Forgetting, without regret, the millions of little weights that rested on his shoulders. And when she raised her eyes to meet his, those rich, emerald eyes... *Home.* He leaned closer and rested his forehead on hers, breathing deeply. *Home.*

"Theodora," he whispered, her name full of warning. "There is no good that will come of this." He pulled back, lowering himself to meet her stare, hands on either side of her. "Let me take care of the king, and you can go off and live your life, however you wish."

"Nice try, Maddox, but I'm not going to live in your debt." She closed her arms around his middle and rested her head against his chest. The last of the flares sporadically filled the night.

"I will always find you, Maddox. In this life and the next."

"How charming."

She looked up at him, moving her arms up his body to wrap them around his neck and pulling him to her, into her kiss.

Plan

Theodora's gut was a tangle of nerves from anxiousness and excitement. Tomorrow she would be at the castle, the cumulation of the past few days coming to an end. While she knew the difficult part still lay ahead, it was euphoric.

She had stayed on the rooftop with Maddox well after the flares had ended and had watched dawn break through the barricade of trees, observing other Lumens setting off their own flares before wobbling to their homes. She and Maddox spoke only occasionally, enjoying the silence. Yet trepidation settled inside her, like a snake coiled and prepared to strike.

Now in the morning light, she strode through the market. The atmosphere was a dull thrum as the elation from the past couple days of festivities faded. Lumens talked amongst each other in a low mumble, sleep wrapped around their words, as if begging for a few more minutes of rest. Children clung to their parents, their antics and games long since forgotten.

The smell of coffee assaulted her nose when she entered Digere. Most of the patrons were silent, lost in their thoughts or ill feelings from the night before. Some rested their chins in their hands and closed their eyes, while others swirled their coffee slowly with spoons.

Theodora gave Albani a nod before taking the stairs to the second floor. The creaking of the steps mimicked her hesitation as her thoughts snagged on the notion of both sides facing each other once more. Honestly, Theodora didn't know which she belonged to.

Did it even matter now?

She banished the thoughts from her mind as she knocked on the door. The Satelle with red hair from the capital party opened it, revealing a storage room stacked with boxes on all sides of the room. Stationed in the middle were five individuals, producing a feeling of claustrophobia Theodora had never experienced before. To her right were the Lumens: the Satelle who'd opened the door, Danika, and Rajveer. Theodora let her gaze stay on Rajveer's a moment longer, watching him, the connection between them now much deeper than previously apparent. On her other side were the Seclusians: Valix and Maddox. The twins Theodora had expected with them were not present.

The tension was like another person in the room, poisoning the air, tainting their lungs as it seeped and weighed heavily amongst them. A collective breath was held as they wondered who would be the first to speak, to break the seal that had formed when the door clicked shut behind her.

Danika cleared her throat, tugging at the collar of her jacket. She looked at Theodora first before she addressed the rest of the room. "The king is aware of the dinner tomorrow. He has also been made aware of the guest that will be appearing." Danika looked nervous and hesitated before continuing, "He expressed his curiosity about meeting you. Although he made some comments about last night, I'm not sure what…"

"Theodora visited Rajveer at the castle," Maddox intervened. Rajveer's eyes caught Theodora's attention; the question clear across his face: *You told him?*

Valix continued instead. "She also took a dip in the lake with your dear prince, which the king witnessed. I guess weaving the story of a possible courtship has already started."

Theodora was stunned. The roles shifted now, as it was Theodora's turn to stare down Maddox so she could silently interrogate *him*. Maddox did not meet her stare, but Valix did with a vicious smile. She rolled her eyes with the realization of what they were attempting to do to the Lumens: disarm them, not with any physical weapons, but with their demeanor.

"Unfortunately, this room isn't big enough for your egos," Theodora countered. "Whether you favor each other or not, we are here for one task. Put your differences aside for a moment. Danika, please continue."

Valix only scoffed before Danika spoke again.

"The dinner will take place in the standard dining hall, which is close to the throne room. It isn't uncommon for Prince Rajveer to present guests there, so nothing should be out of the ordinary. Maybe after this little lake escapade, King Richard will think this is a far more serious relationship. Amicus will enter the room from the hallway, which will be Theodora's cue as I will have locked the secondary exit that leads to the kitchen. The Satelles…"

"I'm sorry to interrupt," Rajveer interjected this time, his intense stare burning into Maddox and Valix. "But why are the two of you here anyway? Your involvement ended once Theodora agreed to the assassination. This plan involves Theodora, and Theodora alone."

"No," Maddox began, "your failed plan involves Theodora and her alone. Once the delightful Danika finishes speaking, I'll let you know the correct course of action we will be proceeding with."

"Well go ahead, mighty Lawless."

"Seclusian. Only *you* see us as Lawless," Valix interrupted.

Rajveer rolled his eyes and folded his arms across his chest.

Maddox continued, "You are making assumptions that the Satelles are going to join in on your assassination of the king. It won't happen. There are those still loyal to your father, whether you want to believe it or not. Four Seclusians, including the two of us, will be at the castle to dispatch any rogue Satelles."

"Should we mark the Satelles who have already pledged their loyalty to Prince Rajveer?" Danika asked.

"If you find it necessary, but it will be obvious which ones are against you. They'll be the ones swinging their blades at you." After a pause, Maddox continued, "Theodora will take care of your king. As I said, the Seclusians will take care of the Satelles."

"And what are we doing?" Danika asked again.

"Stay out of the way and don't fuck it up."

Danika's face flushed red. Even the quiet Satelle in the corner stiffened, his knuckles turning white where he squeezed the hilt of his blade.

"Fine," Danika spat out. She wiped her mouth with the backside of her hand. "Any questions, or can we be done here?"

Valix raised a loose hand in the air. "I have one."

"No one in this room is at all surprised," Rajveer countered.

"Why aren't we being fed?"

"Since there are no other prudent questions, we will see you all tomorrow night. Erm—" Danika turned to face her prince, keeping part of her attention on the Seclusians, no doubt. "How do you want us to exit? We can't all waltz down the stairs together."

"There isn't any appropriate music for waltzing," Maddox interrupted, "so, we will leave the same way we came in."

"And how's that exactly?"

"The window."

The Lumens initially turned to look at the main window behind them, and Theodora had to rein in her smile. Maddox and Valix opened the door and crossed the hall to another room, their footsteps avoiding every creak in the floorboards. Theodora trailed behind the Satelles while they followed curiously to watch from the doorway, as if it were a grand adventure to be found in storybooks.

Valix jumped through the window first, disappearing below the frame. Maddox turned to face them, the alley behind Digere his backdrop. "Long live with king," he said with a wink before falling away.

"He is honestly the most arrogant bastard I've ever met," Danika announced as Rajveer and his other Satelle headed down the stairs.

Theodora halted at the top and watched their descent, uncertain which way to exit, struggling between being a born Lumen but practically raised as a Seclusian. But she was about to murder the king. Whether he was a terrible man or not, she would take his life. Martyr or hero, she would forever be known as the King-Slayer.

She walked into the vacant room, taking in the view of the boxes and crates stored here. As she swiped an orange from an open crate, she slid her body through the window and dropped to the ground below.

Guilty

Maddox's hand gripped the pocket watch inside his pocket, pulling it free from its fabric prison, and spun the turners to the appropriate destination. The portal opened, the electric current radiating before him. He stepped through, leaving his room at the Previt in Seclus and entering an abandoned shop on the western edge of the capital.

He crept out of the shop, quickly making it down the six stairs to the cobblestones before someone took notice of where he came from. His gait allowed him to cover ground quickly to the market square, where he hoped to find Theodora.

As if his thoughts alone conjured her, he detected her stopped at the corner, taking a moment to watch as she fought with her hair to get it wrapped on the top of her head. He continued, and when Theodora noticed him, she crossed her arms over her chest, giving him a knowing smile in acknowledgment. "Of course, you'd find me."

"I knew you wouldn't be able to sit around and count the moments until tomorrow."

She shrugged. "I tried. I took a nap after our long evening and even read a book, but no matter what, I couldn't stop my mind from going to tomorrow's task, trying to calculate the endless scenarios."

A few Lumens raced between the vendors who remained with the setting of Fiedel, the sky around its partial orb a stunning swirl of purples and reds. The lampposts along the walkways and above shops sprouted to life.

"Where are you headed?"

"Nowhere in particular. But I couldn't sit all night." She bounced back and forth on her feet, her need to do something becoming a physical manifestation. "I was going to make some rounds and see if there was anything I could do."

"For klaud?"

"No. More like to help."

Theodora strolled at an unhurried pace, and Maddox followed her lead. She was an enigma to him. She performed wicked tasks, murdered and took the lives of others, and for what purpose? She burned herself ragged for these Lumens and expected nothing in return. Was there some deeper emotional reason? Most likely. Emotional reasoning was not in Maddox's repertoire of understanding. But then why did she perform Lawless tasks at all? Why not pledge full loyalty to Lume?

As she got closer to a row of shops, he witnessed whirls of colorful fabric in a window. The shop of a tailor no doubt. The sign above the door confirmed Theodora was here to see Earleen.

Maddox glanced around the square, noting the few Lumens who ducked their heads and hurried off before he detected the three Satelles headed in their direction. He watched as the probable leader hushed his companions before all three advanced. Maddox's desire to keep his attention on the Satelles disintegrated when he heard shouts from the tailor's shop.

From his peripheral, he saw Theodora halt outside the entryway. The Satelles in the opposite direction slowed their pace as well.

A colossal man with a round face and long ears filled the doorway as he pulled another smaller man along, one Maddox assumed was Earleen. Earleen fought against the grasp and the larger man whirled back, punching Earleen in his large, round spectacles, breaking them off and striking his hawk-like nose. Maddox didn't hear the crack, but at this distance, he could see it was crooked, and presumed Earleen's nose was broken.

Maddox felt as if he was on a tight leash. The Satelles lazily wandered in their direction, and he detected the way Theodora's stance stiffened. The larger man pounded his hand against Earleen's face again as the smaller man struggled to hold himself onto the wooden archway of the door.

The large man stopped his walloping, noticing his audience. Earleen was barely standing, the burly man gripping him by the shirt collar, which was covered in sweat, as blood tangled into the small waves of his hair and dripped down his face. Dry, wheezing breaths rattled from Earleen's chest.

Maddox forced himself to heel, avoiding sudden movements in any direction. The last thing he and Theodora needed right now was to get arrested. They couldn't afford mistakes.

The Satelles directed their attention to Theodora, and for a brief—very brief—moment, Maddox was hopeful for her, until their small smiles turned cruel. "Look, it's the prince's whore."

"Will you do something, please?" Theodora's voice wavered, and it took every ounce of Maddox's control to not react to that sound. Her agony and pain lanced through him.

The Satelles looked at the brute before facing Theodora again. One said, "We see nothing."

The remark was enough to affirm that these weren't Satelles pledged to Rajveer. It was also affirmation that Theodora and Maddox could do nothing to help Earleen, at least not without the Satelles' approval.

With the briefest of nods from the Satelle who spoke, Earleen's assailant revealed a toothy grin. The Satelles started walking away, throwing glances behind themselves to make sure neither Maddox nor Theodora did anything as the man continued to rally his fist against Earleen's face.

It felt like an eternity as they watched the Satelles take their time before disappearing around the curve in the road. The Satelles had to have known Theodora and Maddox would attempt to help Earleen, but they'd used their power to prevent them from doing anything until it was possibly too late. Clearly, Earleen's assailant had assumed that due to his size, he and Theodora would abandon Earleen and walked away.

The man had calculated incorrectly. Theodora immediately raced for Earleen, and Maddox went for the other man.

And Maddox would make sure it cost him.

He punched forward, striking the attacker in his throat. The man reached for his windpipe and attempted to cough down air. Maddox got closer and slammed his palms over the man's ears. The man leaned forward in pain, trying to bring his hands up, when Maddox grabbed his face and pulled down at the same time he jerked his knee upward, slamming the man's face into his knee.

The man's anger from the continuous blows forced him upright, his face now red with frustration and blood. He barreled at Maddox, arms flailing in a whirlwind of directions. Maddox dodged and skirted the attacks, dropping punches to his opponent's torso when allowed the opportunity.

Maddox continued backwards, taunting the man to keep his progression forward, out of Earleen's shop, away from the other buildings. He headed for the middle of the square, occasionally letting strikes hit, keeping the man fueled with irritation.

Tripping over Maddox's foot, the oaf stumbled forward, his face plunging into the water of the fountain. Maddox pressed both of his hands to the back of the man's head, keeping him under. The cold water splashed his jacket, soaking his tunic underneath. Maddox focused on the man's movements, his struggles to find stone to grab onto. Finally, the man attempted to let out a gargled cry, but it was drowned out beneath the surface. The water took the opportunity to fill his already struggling lungs as the last of his life twitched out.

Maddox's pulse was a dull metronome in his ear when he heard Theodora approach. He didn't look at her, but rather kept his attention on how the water of the fountain fell to its impending doom. Maddox shoved the man's body completely into the fountain, letting his lifeless form start its float around the pool.

"Villainous," she acknowledged.

He wasn't sure if her voice broke apart at the end of the word or if he had merely imagined it.

"Effective. Maybe a little inhumane, but I don't care."

Maddox looked at Theodora and noticed the tears coating the edges of her green eyes, the way they searched his and threatened to break him apart right then.

"How do you not feel guilty?" she finally whispered.

"What is there to feel guilty about? He needed to be punished, not only for this attack, but for all the future ones he would have made as well. An easy equation."

"But still, to take a life?"

"It's simple. Humanity is separate from the equation. And when you can see it that way, there is no reason to feel guilt."

They took a moment, listening to the empty sounds around them. The faint laughter down the street, the bangs as tavern doors struck home against their thresholds. Water splashed in the fountain as it cascaded into the pool, some with a different tone now as it struck the corpse of a man floating within.

Maddox finally spoke. "Is he dead?"

"Yes," she replied as she looked back to the shop. The light from within illuminated Earleen's body, where it lay over the threshold.

Maddox wrapped his fingers around her chin, forcing himself to look at her, gazing into her eyes to further prove they were warranted in killing Earleen's attacker. But no tears shone in her eyes now. He could tell, justified or not, that every death broke her apart. Was the same true for the Lawless she killed for him?

She had lost someone else, another friend. Although he could not fix it now, he found solace in the fact that tomorrow, she would be freed of this life.

Culmination

Rajveer paced the eastern and western galleries, stopping before the windows and open door, impatient for her arrival. His father, having left moments ago, insisted on sitting on the covered terrace so he could enjoy the weather during their brief waiting period.

It bothered Rajveer, how calm his father was. But then again, why would he find anything different about this day?

The days counting down to this moment had slowed tremendously. Yesterday, Rajveer had become so restless he'd forced Amicus to train with him to keep his hands busy and his mind entertained. But no matter how many times he wielded his stelgladio, it did not stop him from pouring drinks to drown his emotions and lose hours.

Now he stood here waiting for *her*.

After not assisting her during the storm as it destroyed the festival, he had waited for her after their last meeting with the Lawless, had looked behind him and his Satelles at the bottom of the stairs of Digere. Yet she hadn't followed. Or at least, if she had, it hadn't been immediate.

And when he found out that she had told Maddox about their evening together? He wasn't sure if he was surprised that she'd fulfilled his expectations of her status as a Lawless, or pissed that she'd told Maddox of the events. It wasn't any sort of jealousy, but maybe he'd felt hurt by her disloyalty?

Faint hoofbeats thudded in the distance, echoing his own heart as it raced within his chest. Moments later, the castle gate opened outward to allow him a glimpse of Theodora and the two Satelles following her. The sleeves of her ebony tunic were filled with air, making her appear larger, until her mare stopped, and the fabric collapsed back down to her form. The mare snorted, forcing Rajveer to realize that she rode with no reins, no saddle.

He descended the castle steps, four Satelles close on his heels, and stepped onto the grounds.

Theodora slid down the body of the horse, falling forward gracefully in a dramatic curtsy, tilting her face upward to brandish him a sly wink. "My prince."

He bowed, forcing himself to swallow down his turmoil, which threatened to make him nauseous. He dragged his gaze up her body and noted her corseted skirt, the twin coloring of her tunic. Her choice in apparel was a mockery to the king and the crown, her statement apparent in the blend of two cultures decorating her: Lume, meet Seclus.

"Please allow my Satelles to escort you in. I'll be following shortly after."

Rajveer gestured to Amicus, and Theodora wrapped her arm around his, their movements like ones he had done countless times before over numerous petriks. The other members of the guard proceeded to attempt to approach the mare before they were stopped by Theodora's raised hand.

"My gracious prince, would it be suitable for my horse to merely wander the castle grounds? She does not typically get restrained, as you can tell by the lack of saddle. I promise she won't be a nuisance."

"Consider it done," Rajveer commanded, the Satelles dropping the rope from the horse. He watched Theodora enter the castle, tugging his pack of cigarettes and lighter free from his jacket.

His hands betrayed his nerves, shaking like leaves caught in a windstorm, as he attempted again and again to strike the lighter. Miles brushed close and whispered low in his ear, "What are you doing?"

"The same thing I always do," Rajveer half-shouted.

"This is not the same thing! You don't smoke on the castle grounds. And if my memory is correct, it should be *you* escorting her into the castle, not Amicus."

Rajveer dug out his courage and stared Miles down as he brought in a long pull, the paper and leaves burning quickly and brightly until there was barely anything left. Blowing out the smoke, he watched it swirl about the two of them, the chilled breeze coming in to sweep it away.

Rajveer crushed the end of the cigarette with his hand, the ash coating his metallic fingertips. In a short time, he would be king. He would be crowned and allowed to smoke anywhere he wanted.

Rajveer glared at Miles before facing the castle halls, his and his Satelles' boots announcing their arrival, their grandeur on large display as they entered the dining room.

His father and Theodora were standing in the corner of the room, facing him as he entered, their glasses falling from their lips. Whatever anger had filled him left almost instantaneously as Theodora lifted a smile in his direction. This small and confident woman held his future in her hands, and if she was victorious, he would owe her everything.

"We were talking about what Theo—..." His father hesitated a moment, looking at her to silently inquire about a nickname.

"Theodora," she answered, turning to face the king more directly.

"Yes, alright, Theodora was telling me about what she does around the capital. Rather intriguing, I would imagine. Wouldn't you say so, son?"

Rajveer had to consciously remind his face not to betray him. What could she have possibly told the king? He moved closer to the two of them, and Theodora reached down to the bar to grab him a glass. He took it gratefully and downed a gulp, gathering his courage. How did she do this so effortlessly?

"Yes, well, I find everything about her intriguing."

"Fates tell us, what is it that she does, my son?"

The implications. Warnings and threats felt woven into his unspoken words. Rajveer forced himself to remain calm, but his mind chanted *He knows, he knows* persistently. He drained his glass, and as the wine bubbles popped in his mouth, he looked at Theodora, racing through the conversations they had had with one another. There were too few to provide him with any decent amount of insight as to how she might redirect the questioning.

"She is a peddler. I mean, I don't know what name she gave for herself, but she helps other shopkeepers with their deliveries and assists her fellow Lumens."

His father sipped on his wine, and Rajveer was left in agony as he wondered whether this was what they had talked about. Theodora's face gave away nothing.

Servants began filling the room, the scent of the first course taunting and teasing him, but Rajveer was reluctant to turn to the table without the satisfaction of knowing whether he had answered correctly.

His father proceeded to the table and pulled out a chair, gesturing for Theodora to take it, but she shook her head. "May I please use the washroom first? I don't want to have to interrupt the meal."

"Certainly. Miles, escort Theodora, please."

Theodora curtsied before the king and followed Miles out of the room.

It might have been entirely natural. Maybe Theodora did need to use the washroom, but his mind wouldn't relinquish the endless possibilities of betrayal. He glanced in the direction of his father, only to have his gaze returned with an indifferent expression. Rajveer turned away immediately for fear he might betray himself.

∴ ∵ ∴

Theodora followed Miles back from the washroom as he chose a different route for their return to the dining room. She silently lagged, with no desire to make conversation with the Satelle, especially with her impending departure from the capital.

The curtains framing the dining room doorway came into view, and Theodora glanced at a small hallway table, noticing the arranged flowers and...

She had to force her feet not to stop, to not reach out to touch the mask laid on the tabletop, as if it were another piece of decor. The black mask, wrapped in gears and tubing, brought back her memories. The moment of her parents' death flashed before her eyes, and she was lost to the past until she heard her name, pulling her out.

These masks belonged to the castle.

Now she would be forced to stare at this man, this *prince*, before her. Her emotions were running wild, and she struggled to force them down into submission. But she had seen the mask, as real as everything else in this world. There was no denying what part Lume had played in all of this.

Her breath wobbled as she sucked down gulps of air before entering the dining room once more.

∴ ∵ ∴

The dagger strapped onto Maddox's waist next to his shineguns seemed useless. How Theodora ever found comfort in this type of weaponry was astounding to him.

He stood near the front of the cavern, the Gems patiently awaiting instructions on how to proceed, with two dozen other Seclusians nearby. A couple of the younger ones moved with endless energy, either pacing the length of the cavern, bouncing on their toes, or stretching out their legs and backs with help from the cavern walls. Most conversed softly with their fellows. What they discussed Maddox did not know.

Valix's silhouette approached from an adjoining dimly lit tunnel.

"Stop for some vocatus on the way?" Maddox asked as he took the handful of stelgladios Valix grappled in his hands and passed them to the Gems.

"Does it look like I would have been able to drink with these?"

"No, usually you use a cup for drinking, not a blade."

Valix only glared, swiping his hands down the front of his jacket, dusting away any possible dirt. "Those were all I could find quickly."

"This is adequate. We will be able to get more once we arrive at the castle."

"You still want to enter the training room?"

"That's the plan, Valix. Please try to keep up."

"Yes, sir," was his only reply as he again plucked away a speck of dirt from his shoulder.

Maddox faced the group of Seclusians, dispersing stelgladios among them. "Think logically and stick with the plan." A variety of acknowledgments were gestured or spoken as Maddox pulled out his pocket watch again. "Do not balk at our newest advancement, one that will remove the Lumen king from his throne. Today, we change history."

The portal opened before him, and awes of wonder were barely audible over the swell of current as Maddox stepped through. His boots padded onto the stone of the castle floor, stopping outside where the portal opened, searching the room for any threats. Seeing none, he gestured for everyone else to join.

The Seclusians followed as the door to the training room opened and Danika strode inside, initially oblivious to their presence. When she did notice the audience present, she halted, the portal still yawning open behind him.

"What the fuck," was all Danika could whisper, her eyes wide and her mouth dropped in a gape.

"Danika, lead us to the prince," Valix instructed

"The prince? Oh, right. Right..." She trailed off before speaking directly to Maddox. "Does Theodora know about the—the— She stammered, trying to find the right word. "Tunnel-thing?"

The last of the Seclusians strode through, taking up positions around the room, and the portal pulsed and closed, disappearing completely.

Maddox saw Valix smirk at Danika as he chose to ignore the question.

She threw up her hands in mock surrender. "All I'm saying is Theodora is going to lose her shit."

∴ ∵ ∴

Rajveer heard Theodora's boots as they shuffled in the hallway. She finally approached the table, and everyone took their respective places, including Amicus, who was stationed near the entryway as always.

The staff advanced as if on puppet strings, filling glasses and lifting the lids of the various dishes decorating the table: plums stuffed with goat cheese, roasted rabbit legs, wild mushrooms, and dandelion rice. The smell alone seemed to fill his mouth with the foods' herbs and juiciness.

Rajveer slumped into his chair, trying to keep up his usual demeanor for these dinners, but a hard glace from Theodora had him sitting up straight and placing a cloth along his lap.

"At least it's apparent someone at the table has manners," his father acknowledged before engaging Theodora in conversation. He started with discussions of her parents, who Rajveer hadn't known were tinkerers, or that they were murdered during the dead of night. But every time Rajveer attempted to interject, Theodora was distant.

Was this another ruse? Was the plan to remove *him* from the throne instead of his father? Unwarranted thoughts and ideas invaded his mind, making him question the past few days.

The staff brought another dish forward, placing it before them, featuring fried oranges garnished with sweet vanilla cream. Rajveer leaned forward to sample a bite, annoyed by Theodora and his father's continued discussion. It was another evening of banter getting under his skin, like all the other dinners before, except at the end of *those*, there was little to no aftermath.

He looked to Theodora, beseeching her.

But that little cottage is entirely true.

And yet, assassinating either his father or himself could still get her that.

Rajveer heard a slight tap and watched as another Satelle whispered to Amicus. Amicus entered further into the room, his locks pulled back and held in place with a deep green cord, and strode to the king, relaying the same whisper. Without any further warning, his father rose from the table, throwing his napkin next to his plate.

"You'll have to excuse me, Theodora." His father continued with some smart remark at Rajveer. He only assumed it was something to do with his inadequacies, but Rajveer failed to hear him over the blood pounding in his ears. Everything was going against the plan they'd set in place. First, Theodora. And now, Amicus? His future, his *people,* were at the mercy of some terribly laid out plan by some Lawless.

He'd taken a gamble with Theodora, and it was breaking apart before him, by people he had called friends.

∴ ∵ ∴

Theodora assumed the king had sensed the strain that was forming, but he must have been eager to have someone with whom to discuss Rajveer's faults. Although Amicus whispering into the king's ear was not part of the scheme as they'd discussed yesterday, Theodora knew how to improvise.

She didn't care about defective plans or uncoordinated Satelles, or even spoiled princes. She couldn't care.

The door clicked closed behind the king, Amicus leading the way, leaving Miles and some unknown staff to witness her as she faced down Rajveer.

"What is wrong with you, Theodora? You know this will ruin everything!" He stopped a moment, too flustered for words as he rose from the table. He wiped his fingers across his forehead, flabbergasted at the new situation.

"I don't give a fuck about your plans. Clearly you have had some hidden agenda from the start. If you are going to invite the daughter of the parents you murdered to your palace, you should probably make the evidence less conspicuous."

Rajveer's brows furrowed. "What do you mean?"

She threw the mask she had concealed under her skirt onto the table. The thud as it hit the wood reverberated within her bones, dragging the memories forward from the darkest of places.

"I—" he stammered. She watched as he licked his lips, the bob in his neck as he swallowed. "I don't know what that is."

"How do you not know, Rajveer? When I returned from the washroom, I saw it on one of the tables near the throne room."

"What were you doing near the throne room?"

"That's not the point!"

"That is the point! Why did you need to wander the castle?"

"Why did Lume have my parents murdered?!"

"We didn't!"

And although his voice matched every crescendo in her own thoughts, she didn't care. She never should have taken on this task. She'd spent her life weary of the king, of Satelles, and she'd let herself trust some of them.

Never again.

Alive

Sounds of fighting jolted into the dining room, shinegun blasts ricocheting off stelgladio blades. Rajveer leaned forward to investigate, seeing Theodora slink to the opposite corner of the room in his peripheral as Miles pushed him back against the wall. Members of the king's guard rushed into the room, Miles quickly drawing his own blade as they aimed theirs in the direction of Rajveer.

"Traitor!" Cora, one of the guards, yelled and pointed to Rajveer. "Back away, Miles. He is a traitor to the crown."

Miles remained in his defensive stance. "What are you talking about?!"

"Lawless are attacking. They are swarming the hallways and they say Rajveer let them in." She turned to Rajveer now. "Prince Rajveer Klauduisz, you are under arrest for the attempted assassination of the king and for your treacherous actions against the crown. You're coming with me."

"This is absurd." Miles' voice rose. "You're going to believe Lawless, who are rushing in to attack us, about this? Don't you think they are trying to distract you?"

Cora's face seemed to show she was thinking on this when Amicus ran into the room. "What in the fates are you doing? We need to get the prince to a secure location."

Cora responded, "The Lawless said the prince was planning to assassinate the king and—"

Her words were cut off as Lawless appeared in the doorway. The Satelles in front of Rajveer, who had threatened them with their blades, now directed them in the opposite direction. These Lawless weren't dressed like the one he'd seen during the fire, and it clicked. The mask Theodora had tossed on the table had appeared familiar; he had seen it on the Lawless he'd fought before. He needed to tell Theodora, but when he looked to where she had tucked herself against the wall, she slipped out through the other door.

∴ ∵ ∴

Theodora moved closer to the alternate exit the king had departed from before the Satelles hurried into the room, threatening their own prince. She yanked her dagger from her sheath and crammed herself protectively against a wall. Satelles were accusing Rajveer of being a traitor, and when the Lawless showed up in the doorway? She had to admit the plan had officially turned to shit.

Using the chaos as a distraction, she ducked into the hallway, shutting the door quietly behind her as she searched for the throne room. She followed the hall, which led her to the kitchen. Dodging past cooks and other shocked kitchen staff, she left quickly.

Another hall and multiple turns later, she finally slipped beyond the illustrious pillars rising above the high throne room archways, presenting the grandness of the room within. Theodora knew she would have limited time until the other Satelles noticed her absence and realized the possibility of her being a threat wandering the castle halls.

Adrenaline pumped through her racing heart as she watched the four Satelles standing guard before the throne, protecting the king.

"Come in, Theodora." He waved her forward. "Feel free to shut the door behind you; it might buy us a few extra minutes."

What? Did he realize why she was here? Should she switch up the plan, and abandon this ridiculous idea? But it didn't matter anymore. She had nothing left here. Except for maybe Maddox…

No, she would worry about him when this was over.

Theodora hesitantly proceeded down the long carpet leading to the throne. She watched the king wearily while keeping her attention on the four Satelles. The king downed the last of his wine, setting the glass on an elaborate gold table nearby. The setting light of Fiedel reflected off the lake, illuminating the silhouette of the king and his throne in a brilliant orange glow.

"At long last," the king spoke, spreading his arms wide as he settled deeply into the throne. Not an ounce of concern showed across his face. She didn't speak. She assumed that he would want the floor to remain his, and she had guessed correctly. "I have been waiting for this moment, you know? Far longer than you could even begin to imagine."

∴ ∵ ∴

Shots were fired by the Lawless, and Rajveer ducked behind one of the dining room chairs. The Satelles swung their blades in an attempt to deflect the blasts, but with the increased firing, chunks of wood splinted off the table. The blasts were creating holes in the wall.

At least, Rajveer thought humorously, maybe this would mean no more pointless and elaborate meals would be expected soon.

Rajveer spotted a Lawless as he crept the length of the table. Remaining crouched and making it appear as if he hadn't yet noticed him, Rajveer shifted his hand to his waist, where his blade had remained sheathed.

The Lawless fired and Rajveer slid the blade out, swiping the charge away. It bounced into the chandelier. Crystals rained around the room, adding to the anarchy. Rajveer strode forward, deliberately closing the distance between him and his attacker. Fire, swipe, fire, swipe, until furniture no longer blocked his path. He charged, swinging the blade up in anticipation of his calculated blow.

As his stelgladio dropped for the Lawless' head, the Lawless brought his shinegun upward, stopping Rajveer's momentum on the trigger guard of the shinegun.

Rajveer weighed his options on how to counter the block. He could continue with added force, or ram his head into the Lawless' face.

The Lawless shifted, pulling another shinegun from where it was holstered around his waist. He aimed over their tangled weapons at Rajveer's head. On instinct, Rajveer grabbed the end of the barrel with his metal hand.

He was out of breath, running out of ideas, with his hand wrapped around the end of the shinegun, and was gambling on whether it would withstand the charge or not.

Another barrel of a shinegun pointed at the Lawless' head. Rajveer turned to identify who it was. His eyes followed the arm of a golden jacket and traced over the cape as it draped down a shoulder, where dreads of browns and blacks twinned around each other. Maybe Amicus merely hoped to deceive the Lawless, but most knew shineguns had to be registered. There was no way Amicus would be able to fire the weapon and kill his opponent.

But when the charge exploded through the Lawless' head from the shinegun Amicus held, Rajveer could barely process the sense of betrayal he felt, because his shoulder blade was engulfed in agony. And he had no ability to do anything but fall to the floor with the dead Lawless as everything went black.

∴ ∵ ∴

"Why are there so many of them?" Valix began firing his shinegun again as four Satelles stepped out of their hiding places, attempting to rush them. "It's like they multiplied."

"We need to get to the throne room," Maddox said.

"It's not like I'm doing nothing here."

"Let's go!" Maddox bellowed. "We need to find Theodora now! We're running out of time."

∴ ∵ ∴

Theodora watched the Satelles leave the throne room on the king's order. She felt her brows pinch together in confusion over why he would release his guards, but she was focused more on his prior comment about anticipating this moment.

She used that to precede her questioning instead. "Did you know?"

"Know what? Who killed your parents? At the time, no. But I was informed about who it was later, and since then, I traded large amounts of klaud to learn secrets about you. And here you are to finally grant my wishes." He waved ceremoniously outward, as if the room was filled with his citizens, people he had to persuade with his tales of grandeur, when it was only her. "Come on up, Theodora. Kill me and let Lume reign."

"Why are you so eager to die?"

"Well, didn't he tell you the final part to all this?"

"Who? Rajveer? No, I only saw the mask sitting on the table. The same mask that I *know* was worn by my parents' murderers."

The king chuckled. "It wasn't Rajveer. He doesn't have enough cleverness to make plans fiedations in advance. I'm sorry, but think about it. Who does that mask belong to?" He paused as if he honestly assumed she would be able to figure it out. "Do you need a hint?" The smile on his face was greedy and vicious.

"Why?"

"I needed my son out of the way. Lume could never survive under his rule. He can't follow a simple tradition of marrying the appropriate person. How could he manage a capital? So, I reached out to the Seclusians. And with a little patience, my name will continue in infamy in the storybooks, and Lume will triumph."

"You're a liar!"

"Am I? Oh, my sweet child. You know so little."

As much as she wanted to understand, as much as she had thousands of unanswered questions swirling in her mind, her emotions overtook all logic. But that was always her way. Death might have taken her parents from this world, on the orders of Lume, its pathetic king and spoiled prince.

As she walked up the final steps of the throne, she looked down at him. She sucked in a breath before slashing her blade across the king's neck. She watched the warmth of his espresso eyes die away, and in the first moment in far too long, she finally felt alive.

Lume

The blood on her hand was warm as she dropped the dagger in the dead king's lap, but with her whirlwind of emotions, she barely registered how it stuck to her fingers. All she felt was relief. She couldn't care about what happened to Rajveer, or to Lume even. Anybody could take the mantle at this point. All she needed—no, all she wanted, was to be free of this place.

She reached for the crown, wrestling it from the tangles of the king's dark peppered hair. The crown was simple in its beauty. Gold swirls wrapped in on themselves, with emeralds folded into the metal catching the light when it moved. Something so simple held so much power. It had brought underground cities from their darkened tunnels, caused rebellions and chaos, and made sons murder their fathers.

That kind of power, that kind of responsibility, that kind of control... and for what?

She dropped the crown to the ground, where it tumbled at the king's feet. Turning from his lifeless body, she made her way back down the steps, the crimson carpet meeting her boots in mockery. She overheard boots on tile and jerked her head upwards, forcing herself to stop short when Maddox entered the throne room.

In his hand he held the head of a Satelle, thankfully one unknown to her, but it didn't squelch the betrayal seeping into her veins. Blood dripped from where his neck used to be and his eyes were rolled back, a scream held permanently on his lips. Theodora's hand moved to her mouth to hold back the nausea that threatened her as Valix and the twins came in behind him.

Grief rolled forward, taking the place of the nausea she'd held deep within her stomach. And when her stomach could hold no more, it spilled into her bones.

Settling there. Hardening her.

She was at a loss for words. She didn't know where to begin, and she didn't know why. Her eyes burned with tears. The only possible explanations were crushing her from within before her mind would even agree to acknowledge them. She raised her eyes to Maddox's dark ones, forcing herself to not let those damn tears fall.

Were the allegations from the dead king true?

"What are you doing?" The words barely left her dried mouth, scraping along her tongue and burning her throat.

Maddox shifted his weight, throwing the Satelles' head to the ground by her feet. It was only then that Theodora noticed the shinegun Maddox held in his other hand. She had no weapon, and she needed something. The muscles in her leg worked as she twitched to stomp her boot.

"Don't," Maddox bellowed as he raised the shinegun in her direction.

Seconds ticked by, Lume holding a collective breath.

"Why?"

She couldn't stop her voice from cracking as it echoed the sound of her heart breaking. She didn't have to explain her question; she knew he would have to know what she meant.

"I'm sorry, charm. I told you, no good would come from this."

She was grasping blindly at the betrayal, but to no avail. She grabbed onto the only emotion she had left: anger.

"So, in all of this, I was just another pawn on your board."

"I wouldn't say pawn. You are far too skilled and valuable for that role. I knew that you had a vested interest in killing the king and the reward that was promised. It kept you focused on what I needed, and blind to what I didn't want you to see."

The air was becoming thick with emotions, heavy with lies.

And her words? Her questions?

Lodged in her throat, uncertain of where to begin.

All she knew was that *this* was not how it was supposed to happen. Hadn't she destroyed her life enough, given enough, to the fates? Maddox was supposed to abolish the evil that tainted the throne, help to inspire change in Lume, find a balance with Seclus... let them be together! Her brain couldn't process through the words Maddox was offering her, and her heart... She never wanted this.

She never wanted to be Lume's hero, or its martyr. She wanted her cottage and freedom. Maybe a chance at love.

And that's what this was. Wasn't it? She had made choices blindly, thinking Maddox was going to change, thinking he was going to choose her above all others. Because wasn't this how all the stories were supposed to end? Through all the gray, and the chaos, and the bullshit, wouldn't her love be enough?

"But I loved you."

The words barely escaped her lips, a whisper in the massive grandness of the throne room, filled with rich walls and opulent flooring, with a spacious throne set before a spectacular view. And there within it was Lawless woman who had fallen in love.

Valix yelled over his shoulder, "The king! She's killed the king." Satelles rushed into the room, filling either side. Valix shouted in their direction. "She murdered the king!" They pointed their stelgladios in her direction.

Valix turned to Maddox. "Kill her, sir."

Theodora looked into Maddox's eyes, pleading for him to stop whatever he intended. She opened her mouth to speak when a pain ignited within her gut and the fire spread outward. She looked down and watched as the leather of her corset darkened, her blood seeping and staining it.

The pain should have been unbearable, but the lies and betrayal she felt was a narcotic, numbing all physical feelings. If only it could numb her heart, too. She slipped down to the floor, but she clung to consciousness and brought her arms out to catch herself. Sitting, she watched as black surged into her vision, and she blinked it back, refusing to give up this one thing she had left to claim as her own.

She wobbled her head upward to look at Maddox and realized he was not watching her anymore, but staring intently at Valix instead.

Valix spoke again. "You hesitated, sir."

You hesitated, sir.

Her world cracked open, and her heart broke apart. But no amount of anger she felt was enough to fight the fire that consumed her as darkness took its claim.

Home

Maddox stared out the windows of the throne room, annoyed with Valix standing next to him. He watched the lights of shineguns and flash crystals illuminated the trees outside as his men fought against the Lumen rebels. The sound was a low rumble through the glass. It reminded him of the flares he had witnessed with Theodora just a few days prior. But instead of these flares being in the sky, they ignited across the ground.

Theodora.

It had only been a few hours since Valix had shot Theodora. After which, Maddox had lost her in the chaos of securing his position over the throne, as the Lumens were without a king or prince. The Seclusians easily took control of both the castle and the capital.

"We're almost finished here on Lume, committer," Valix interrupted his thoughts. "We could speed up the process in attempting to locate the missing guard. Maybe some public hangings?" he continued.

Maddox rolled his eyes. His associate had the excitement of a child. Maddox rejected the idea, but was unsuccessful in keeping the anger from his voice.

He continued to stare at the flashes and attacks before him, entranced by the explosives, but his mind kept wandering back to her. Back to her smile. He had only shown her pieces of who he was, a small fragment of the monster and viciousness that lay within. But instead of backing away in fear, she fell in love with him.

He stormed out of the throne room, Valix quick on his heels, through various rooms he hadn't yet investigated to a covered terrace. The twins rose from where they lounged in overstuffed chairs, and Maddox met their gazes. They nodded in return.

Slipping his hand into his vest pocket, he pulled out the pocket watch, the metal barely warmed by the fabric that housed it. He pushed down on the turner, allowing the metal to spring away from the face of the watch. The newly added inscription surrounded the face, a taunt and reminder: *in this life and the next*. It was a memory of her, now permanently etched into the piece.

Maddox pulled the turner. One click. Two clicks. Three clicks. With a swipe of his thumb, the face rotated out of the traditional setting, opening a portal within the middle of the circle formed by their bodies.

In the sights of the portal lay a different view than airy curtains and alabaster pillars: the world of Petram. Bright green fields and even brighter skies of blue. The warmed wind gusted about the border of the portal, attempting to drift into the chilly terrace.

Valix and the twins raced into the portal, eager to return. Maddox stepped through, the portal rippling across his body as his black shoes met the ground on a world he called home.

to be continued...

Glossary

Astrum: third season of the year where leaves fall for the land mass of Lume

Conlis: the road connecting to this city is less traverse due to travel through the mountains; provides stones, minerals, rocks, and gems in raw form as well as coal and gas to the cities

Dumgun: pistol; metal, typically found with filigree decorating the barrels and wooden grips; uses single metal projectiles for firing; outdated technology having been replaced

Erysum: city known for its farmlands; provides most of the vegetation and crops for the continent; city was established after most of the grounds around the capital were destroyed of their nutrients and resources

Fevron: season after heim where vegetation begins to appear for the land mass of Lume

Flash Crystal: grenade; filled with explosive glass; detonates on a time delay, dispersing glass into the immediate radius, accompanied with bright flash of light which may also case temporary blindness

Freta: port city bringing in most of Lume's fish and other ocean resources; Satelles here continue to take scouting ships into the ocean in an attempt to determine if any other land masses exist

Fiedation: measurement of time using Fiedel; multiple petriks create a fiedation; similar to an Earth year.

Fiedel: Star which is orbited by the planet containing the land mass of Lume; provides light and warmth

Fiedelight: light provided from the star, Fiedel

Heim: coldest season of the year for the land mass of Lume

Lawless: those acting against the king; *syn.* criminal, rebel

Nemaaer: northern city dedicated to serving the fates; due to its farther distance, it is mostly forgotten; the city produces its own resources and is self-sufficient, although Satelles are stationed here to ensure the citizens are compliant to the king

Lume: the only known land mass on the planet containing the capital where the king's castle is located; used interchangeably to mean the capital, the continent, and the planet

Perdit: abandoned city; the king is attempting to remove its story from Lumen history

Petram: a natural satellite of the planet containing the land mass of Lume; reflects some light from Fiedel and is seen mostly at night

Petrik: measurement of time using Petram; multiple days create a petrik; similar to an Earth month

Satelle: king's guard

Satelle Cape: worn strictly by Satelles; normally worn to appear as fabric, but can absorb phaser light projectiles such as those from shineguns; cannot block flash crystal projectiles or metallic projectiles

Seclus: underground city which was recently built; home for Seclusians who are known for their advanced technology and weaponry; most Seclusians are considered Lawless by Lumens

Shinegun: pistol; unknown metal found with black solar charged stonelike pieces along the top; it is charged through star light and fires a blast of hot, radiating light or phasers; weapon is fingerprint activated with a registry held by Seclus

Solta: warmest season of the year for the land mass of Lume

Stelgladio: electrified sword; it creates a magnetic field to warp phaser light projectiles such as those from shineguns; can also be used to block metallic projectiles such as those from a dumgun

Vindem: city known for their grape vines and berries; harvests and cultivates the king's wine hoard; works in tandem with Erysum to create the beer that is exported to the various cities

Vocatus: common spiced liquor with notes of cinnamon only served in Seclus; current ingredients remain unknown; Origin: unknown.

Lume Playlist

Paint It, Black	Ciara
Tortured Soul	Chord Overstreet
Future Heroine	Ecca Vandal
Man or Monster (feat. Zayde Wølf)	Sam Tinnesz
Darkside	Alan Walker
Outsiders	Au/Ra
Everybody Wants to Rule the World	Weezer
Wrecked	Imagine Dragons
Control	Halsey
Walked Through Hell	Anson Seabra
Fell On Black Days	Soundgarden
Down River	The Temper Trap
Scars	Papa Roach
Unsteady	X Ambassadors
We Found Love	Rihanna
Feel Something	Jaymes Young
Flares	The Script
Sinners	Lauren Aquilina
Take On The World	You Me At Six
Run to You	Lea Michele
Gravity	Sara Bareilles
Let's Kill Tonight	Panic! At The Disco
Knights of Cydonia	Muse
Darkside	Oshins
Monster	Beth Crowley
Power	Isak Danielson
Are You With Me	nilu
Don't Let Me Go	RAIGN
Before You Go	Lewis Capaldi
The Light Behind Your Eyes	My Chemical Romance

About the Author

TM Ghent is a fictional tragedy writer who loves grumpy sunshine. She likes to dance the line between morally gray hero and anti-hero characters, forcing you to question who the villain is. Refusing to adhere to standard literary tropes, her stories break the mold and defy genres, such as blending fantasy world-building with sci-fi tech or historical plots in contemporary settings.

TM Ghent lives with her husband and two children on the east coast of the United States and enjoys the early mornings with a hot cuppa.

When she isn't lost in a story, she can be found making music in her local community band, heading to the movies with her husband, or raising her own little word-lovers.

Every story doesn't end with a happily ever after.

You can follow her on Instagram at @tm_ghent

What's Next?

Petram
Lume Book 2

Don't miss the epic conclusion of this duology! To be released 10 November 2023, see the opening chapter in the next pages…

Prince Rajveer Klauduisz woke with a start, but instantly regretted it. He had been stabbed in the back. Most likely figuratively, but the severe pain in his shoulder told him literally as well. He attempted to raise his head from the stone floor, trying to figure out where he was.

"Stay still," someone hissed at him as they pushed his face back into the dirt and dust covering the stone.

Miles.

"It fucking hurts," he responded.

"I'll bet it does, sir. You're bleeding everywhere. But I need to get you patched up so we can leave."

"Leave? Miles." Rajveer's breaths were a large feat. Quick and incumbered. "I can barely hold on as it is. I'm not sure moving is a good idea."

Ignoring him, Rajveer felt Miles' hands dig into his shoulder. "Alright, sir, this is going to hurt."

Rajveer wasn't sure he knew what Miles was talking about until the pain in his shoulder engulfed in flames. He heard Miles gag next to him, and Rajveer's own stomach turned when he felt thread getting pulled through his raw skin.

"Hold on, sir. Help is almost here."

And even though the assault on his back continued, he heard the tapping of, an echo to the pounding pulse in his skull. He was barely holding onto consciousness. There was no way he could lift his head, so he lifted his eyes instead, looking in the direction of the hurried walk.

The sound of the shoes shuffled to a stop, but Rajveer was sure the person wasn't near him yet. He heard clattering and movement, then a woman's voice. "We need to get her back up to the throne room soon."

Miles responded, "Sure, sure. I'm finishing up with the prince now. Can you come over and monitor him so I can make sure she is good to go upstairs?"

A sound of acknowledgement hummed in the air before there was the quick switching of people. He opened his eyes again to see who'd walked in his direction.

Her skin was flawless, the color of the sandy beaches along Freta's coastline. She kneeled next to him, and he focused on her face. Her deep brown eyes were highlighted with long lashes, following the lines of her sculpted brows. Rajveer couldn't see the pout she wore through the fog of his brain, but he could remember what it must look like. Just as he remembered the gold ring tucked around her left nostril.

Rajveer's eyes closed for a second and slowly opened once more. He couldn't see her hair, not because of the dimmed lighting or the long shadows in the room, but rather because of the headscarf wrapped around her head, framing her face.

He had to be dreaming. A collision of truths and lies swirled before him, because Alouette shouldn't be here.